THE NEW HARVEST

THE NEW HARVEST

AGRICULTURAL INNOVATION IN AFRICA

Second Edition

Calestous Juma

OXFORD
UNIVERSITY PRESS

OXFORD
UNIVERSITY PRESS

Oxford University Press is a department of the University of Oxford.
It furthers the University's objective of excellence in research, scholarship,
and education by publishing worldwide.

Oxford   New York
Auckland   Cape Town   Dar es Salaam   Hong Kong   Karachi
Kuala Lumpur   Madrid   Melbourne   Mexico City   Nairobi
New Delhi   Shanghai   Taipei   Toronto

With offices in
Argentina   Austria   Brazil   Chile   Czech Republic   France   Greece
Guatemala   Hungary   Italy   Japan   Poland   Portugal   Singapore
South Korea   Switzerland   Thailand   Turkey   Ukraine   Vietnam

Oxford is a registered trade mark of Oxford University Press
in the UK and certain other countries.

Published in the United States of America by
Oxford University Press
198 Madison Avenue, New York, NY 10016

Library of Congress Cataloging-in-Publication Data
Juma, Calestous, author.
The new harvest : agricultural innovation in Africa / Calestous
Juma. — [Second edition]
p. cm.
Includes bibliographical references and index.
ISBN 978–0–19–023723–3 (pbk. : alk. paper) 1. Agriculture—Economic
aspects—Africa. 2. Agricultural innovations—Economic aspects—Africa.
3. Economic development—Africa. I. Title.
HD9017.A2J86 2015
338.1'6096—dc23
2015003225

1 3 5 7 9 8 6 4 2

Printed in the United States of America on acid-free paper

In memory of Christopher Freeman, father of the field
of science policy and innovation studies

CONTENTS

FOREWORD TO THE SECOND EDITION

The publication of the first edition of *The New Harvest* was written as a manifesto for the optimist. It was meant to be a call to action. The book was released on the heels of a series of food price spikes and the Arab Spring uprising in North Africa. The Arab Spring provided clear evidence that the ability of a country to feed itself was linked to its national security. The events helped create a sense of urgency among African leaders to focus on agriculture as a foundation for the long-term economic transformation of the continent.

This edition serves four purposes. First, it acts as report card on what has been achieved since the release of the first edition. The main message from the lessons of the last five years is that countries can overcome their most intractable challenge if they can bring high-level political capital to bear on the search for solutions. The first edition was launched in Arusha (Tanzania) by the five heads of the East African Community (Burundi, Kenya, Rwanda, Tanzania, and Uganda). The fact that the leaders were willing to launch a book they had not yet read testifies to their willingness to take risks with new ideas to address a persistent problem.

The second purpose is to underscore Africa's latecomer advantages. Exponential growth in science, technology, and engineering is expanding the range of technical knowledge that the continent can marshal for agricultural transformation.

This edition highlights the importance of leapfrogging in agricultural biotechnology. This includes the use of new genetic technologies that do not involve moving genes across species. The same leapfrogging strategies can be applied in other fields such as information and communications technologies, satellite technology, unmanned aerial vehicles, renewable energy, synthetic biology, and polymer chemistry. Judicious use of technologies from these fields can help Africa adopt more ecologically sound agricultural practices.

The third purpose of this edition is to renew its optimistic message that Africa can feed itself in a generation. At its release this message appeared to some observers as wishful thinking. Interestingly, the international press is usually accused of focusing too much on negative news from Africa. In this case the press played a critical role not only in helping to shape the message but also in giving it global currency. The sense of optimism and the rising number of champions of agricultural transformation among African heads of state played a key role in getting world leaders in business and philanthropy to increase their investments throughout the continent.

The final objective is to help sustain the momentum that was created by the first edition. Africa's agricultural transformation will require long-term policy commitments. Much more needs to be done to scale up the various experiments across Africa into long-term strategies. The view that agriculture is just a stepping stone in linear stages of economic development no longer holds true. Future strategies for economic inclusion and sustainability demand a systems approach in which agriculture will remain important.

The future of Africa belongs to its youth. The achievements reported in this book should help young Africans to appreciate the need to keep their eyes set on opportunities for agricultural improvement. As Winston Churchill so aptly put it: "The pessimist sees difficulty in every opportunity. The optimist sees the opportunity in every difficulty."

ACKNOWLEDGMENTS

This revised edition is a product of the Agricultural Innovation in Africa (AIA) project, funded by the Bill and Melinda Gates Foundation. Its production benefited greatly from the input of the African Union High-Level Panel on Science, Technology, and Innovation. We are grateful for the intellectual guidance and written contributions of the members of the original AIA International Advisory Panel. The production of this book would not have been possible without their genuine support for the project and dedication to the cause of improving African agriculture. We especially thank Dr. Akin Adesina (Minister of Agriculture and Rural Development, Nigeria) for his outstanding leadership for promoting agricultural transformation in Africa. The Advisory Panel included John Adeoti (Nigerian Institute of Social and Economic Research), Aggrey Ambali (NEPAD Planning and Coordinating Agency, South Africa), N'Dri Assié-Lumumba (Cornell University, USA), Zhangliang Chen (Guangxi Zhuang Autonomous Region, Nanning, People's Republic of China), Mateja Dermastia (Center of Excellence PoliMaT, Slovenia), Anil Gupta (Society for Research and Initiatives for Sustainable Technologies and Institutions, and Indian Institute of Management, India), Daniel Kammen (University of California, Berkeley), Margaret Kilo (African Development Bank, Tunisia), Hiroyuki Kubota (Japan International Cooperation Agency, Tokyo), Francis

Mangeni (Common Market for Eastern and Southern Africa, Lusaka, Zambia), Magdy Madkour (Ain Shams University, Cairo, Egypt), Venkatesh Narayanamurti (School of Engineering and Applied Sciences, Harvard University), Robert Paarlberg (Wellesley College and Harvard University), Maria Jose Sampaio (Brazilian Agricultural Research Corporation, Brazil), Lindiwe Majele Sibanda (Food, Agriculture and Natural Resources Policy Analysis Network, South Africa), Greet Smets (Biotechnology and Regulatory Specialist, Essen, Belgium), Botlhale Tema (African Creative Connections, Johannesburg, South Africa), Jeff Waage (London International Development Centre), and Judi Wakhungu (African Centre for Technology Studies, Nairobi, Kenya).

We are deeply grateful to the information provided to us for both versions by Juma Mwapachu (Secretary General), Alloys Mutabingwa, Jean Claude Nsengiyumva, Phil Klerruu, Moses Marwa, Flora Msonda, John Mungai, Weggoro Nyamajeje, Henry Obbo, and Richard Owora at the Secretariat of the East African Community (Arusha, Tanzania) on Africa's Regional Economic Communities (RECs). Sam Kanyarukiga, Frank Mugyenyi, Martha Byanyima, Jan Joost Nijhoff, Nalishebo Meebelo, and Angelo Daka (Common Market for Eastern and Southern Africa, Zambia) provided additional contributions on regional perspectives. We have benefited from critical insights from Nina Fedoroff and Andrew Reynolds (US Department of State, Washington, DC), Andrew Daudi (Office of the President, Lilongwe, Malawi), Norman Clark (African Centre for Technology Studies, Nairobi, Kenya), Daniel Dalohoun (AfricaRice, Cotonou, Benin), Andy Hall (UNU-MERIT, Maastricht, the Netherlands), and Peter Wanyama (Mohammed Muigai Advocates, Nairobi, Kenya). Additional information for the book was provided by John Pasek (Iowa State University, USA), Stephanie Hanson and Hilda Paulson (One Acre Fund, Bungoma, Kenya), S. G. Vombatkere (Mysore, India), Akwasi Asamoah (Kwame Nkrumah University of Science and Technology, Kumasi, Ghana),

Jonathan Gressel (Weizmann Institute and Trans-Algae Ltd., Israel), Will Masters (Tufts University, USA), Martha Dolben and Julia Pettengil (African Food and Peace Foundation, USA), C. Ford Runge (University of Minnesota, USA), Mateja Dermastia, Marco Prehn and Mark Moore (AGCO), Jon Vandenheuvel, Kris Klokkenga, and Issa Baluch (Africa Atlantic Franchise Farms, Ghana), Yemi Akinbamijo (FARA, Ghana), Mira Mehta and Shane Kiernan (Tomato Jos, Nigeria), Kola Masha (Babban Gona, Nigeria), Alexander Oppenheim (Vital Capital), Su Kahumbu (iCow, Kenya), Sara Menker (Gro Intelligence, New York), Diane Ragone (Breadfruit Institute, Hawaii), Laura Pereira, and Martin Fregene III (Director, BioCassava Plus, Danforth Center, St. Louis, USA).

We would like to thank our dedicated team of researchers who included Kate Collins, Amandine Lobelle, and Ingrid Ohna at the Harvard Kennedy School. We are grateful to those who provided assistance on the first edition: Edmundo Barros, Daniel Coutinho, Manisha Dookhony, Josh Drake, Emily Janoch, Beth Maclin, Julia Mensah, Shino Saruti, Mahat Somane, Melanie Vant, Yunan Jin, Jeff Solnet, Amaka Uzoh, Caroline Wu, and Greg Durham.

We are particularly grateful to Richard Carey, Kaori Miyamoto, and Fred Gault (OECD Development Centre, Paris). Further input was provided by David Angell (Ministry of Foreign Affairs, Ottawa), Jajeev Chawla (Survey Settlements and Land Records Department, India), David King (International Federation of Agricultural Producers, Paris), Raul Montemayor (Federation of Free Farmers, Manila), Charles Gore (United Nations Conference on Trade and Development, Geneva), Paulo Gomes (Constelor Group, Washington, DC), Ren Wang (Consultative Group on International Agricultural Research, Washington, DC), David Birch (Consult Hyperion, London), Laurens van Veldhutzen (ETC Foundation, the Netherlands), Erika Kraemer-Mbula (Centre for Research in Innovation Management, Brighton University, UK), Khalid El-Harizi (International Fund for Agricultural Development,

Rome), Adrian Ely (University of Sussex, UK), Watu Wamae (The Open University, UK), Andrew Hall (United Nations University, the Netherlands), Rajendra Ranganathan (Utexrwa, Kigali), Alfred Watkins (World Bank, Washington, DC), Eija Pehu (World Bank, Washington, DC), and Wacege Mugua (Safaricom, Ltd., Nairobi).

Our ideas about the role of experiential learning have been enriched by the generosity of José Zaglul and Daniel Sherrard at EARTH University and their readiness to share important lessons from their experience running the world's first "sustainable agriculture university." We drew additional inspiration and encouragement from Alice Amsden (Massachusetts Institute of Technology, USA), Thomas Burke (Harvard Medical School, USA), Gordon Conway (Imperial College, London), David King (University of Oxford, UK), Yee-Cheong Lee (Academy of Sciences, Malaysia), Peter Raven (Missouri Botanical Garden, USA), Ismail Serageldin (Library of Alexandria, Egypt), Gus Speth (Vermont Law School, USA), and Yolanda Kakabadse (WWF International, Switzerland).

We are grateful to the Bill and Melinda Gates Foundation and in particular to Brantley Browning and Rinn Self, for continuing support to the Agricultural Innovation in Africa Project at the Belfer Center for Science and International Affairs of the Harvard Kennedy School. We are also indebted to our colleagues at the Harvard Kennedy School who have given us considerable moral and intellectual support in the implementation of this initiative. We want to single out Graham Allison, Venkatesh Narayanamurti, Bill Clark, Nancy Dickson, and Kevin Ryan for their continuing support.

Special credit goes to Katherine Gordon (AIA Project Coordinator, Harvard Kennedy School), who provided invaluable support and research assistance on the revised edition. We thank Angela Chnapko (Oxford University Press) for her constructive editorial support and guidance. We would also like to thank Walter Lamberson for pointing out citation errors in an earlier version of the text. Finally, we want to thank Issa

Baluch of Africa Atlantic Holdings for his continued support of our work.

New ideas need champions. We would like to commend Sindiso Ngwenya (Secretary General of the Common Market for Eastern and Southern Africa, Lusaka) for taking an early lead to present the ideas contained in the first iteration of this book to African heads of state and government for consideration. As a result of his efforts, African leaders as well as the private sector prioritized agriculture and food security. This revised edition is a testament to the development of African agriculture since 2010.

INTRODUCTION

Albert Einstein said that "an empty stomach is not a good political advisor." Heeding this warning, African leaders have been paying considerable attention to food security. They have also recognized that investing in agriculture contributes to overall economic development and poverty reduction.

The January 2014 summit of the African Union (AU) marked the official launch of the "Year of Agriculture and Food Security in Africa."[1] It also marked the tenth anniversary of the adoption of the Comprehensive Africa Agriculture Development Programme (CAADP). In those 10 years, the 11 countries that met or exceeded their CAADP targets saw agricultural productivity and food security increase.[2] The declaration of 2014 as the Year of Agriculture demonstrates a continued recognition of the importance of agriculture, not only in the context of food security but also as a primary driver of economic development.

The first edition of *The New Harvest* argued that Africa could feed itself in a generation. In the five years since publication, several African presidents have risen to the challenge and have made agriculture a focal point of their development agendas.[3] By 2015 nine countries have joined the $10 billion Grow Africa Partnership. This multi-stakeholder platform was developed in support of CAADP to accelerate agricultural transformation throughout the continent, and it has redefined the role of the private sector in agricultural development.

In his acceptance speech as Chairman of the Assembly of the AU in February 2010, the late President Bingu wa Mutharika of Malawi said:

> One challenge we all face is poverty, hunger and malnutrition of large populations. Therefore achieving food security at the African level should be able to address these problems. Africa is endowed with vast fertile soils, favourable climates, vast water basins and perennial rivers that could be utilized for irrigation farming and lead to the Green Revolution, and mitigate the adverse effects of climate change. We can therefore grow enough food to feed everyone in Africa.[4]

Mutharika's statement laid out a clear vision of how to approach Africa's agricultural challenge. This book builds on this optimistic outlook against a general background of gloom that fails to account for a wide range of success stories across the continent.[5] African agriculture is at the crossroads. Persistent food shortages are now being compounded by new threats arising from climate change. But Africa faces three major opportunities that can help transform its agriculture into a force for economic growth. First, advances in science, technology, and engineering worldwide offer Africa the new tools needed to promote sustainable agriculture. Second, efforts to create regional markets will provide new incentives for agricultural production and trade. Third, a new generation of African leaders in the public and private sectors is helping the continent to focus on long-term economic transformation through entrepreneurship and innovation. This book provides policy-relevant information on how to align science, technology, and engineering missions with regional agricultural development goals.[6]

This book argues that sustaining African economic prosperity will require significant efforts to modernize the continent's economy through the application of science and technology in agriculture. In other words, agriculture must be viewed as a knowledge-based entrepreneurial activity.[7] The argument is based on the premise that smart investments in agriculture will have multiplier effects in many sectors of the economy and will help spread prosperity. More specifically, the book focuses on the importance of boosting support for agricultural research as part of a larger agenda to promote innovation, invest in enabling infrastructure, build human capacity, stimulate entrepreneurship, and improve the governance of innovation.

The emergence of Africa's Regional Economic Communities (RECs) provides a unique opportunity to promote innovation in African agriculture in a more systematic and coordinated way.[8] The Common Market for Eastern and Southern Africa (COMESA) has been the most effective at expanding prospects for prosperity by creating space for economic growth and technological innovation. For example, in 2013 COMESA launched its own research innovation council, an advisory group comprising African experts in science and technology. The creation of this council is a clear acknowledgment by top leadership that science, technology, and innovation can play an important role in the region's development. COMESA aims to close this gap by connecting scientists and engineers; by providing member states with technical knowledge and advice relating to economic development as needed; and by spurring entrepreneurship through the establishment of an annual award and an innovation fund.

This edition builds on the findings of the African Union High-Level Panel on Science, Technology, and Innovation. The panel's main recommendations focus on three pillars: infrastructure, higher technical training, and entrepreneurship. This book aims to provide ideas on how agriculture can spur economic development in Africa by focusing on these three pillars. It outlines the policies and institutional changes needed to promote agricultural innovation in light of changing ecological, economic, and political circumstances in Africa.

This book also explores the role of rapid technological innovation in fostering sustainability, with specific emphasis on sustainable agriculture. It provides illustrations from advances in information technology, mobile technology, biotechnology, and nanotechnology. It builds on recent advances in knowledge on the origin and evolution of technological systems. Agricultural productivity, entrepreneurship, technical training, and value addition foster productivity in rural-based economies. In many poor countries, however, farmers, small- and medium-sized enterprises, and research centers do not interact in ways that accelerate the move beyond low value-added subsistence sustainable agriculture. Strengthening rural innovation systems, developing effective clusters that can add value to unprocessed raw materials, and promoting value chains across such diverse sectors as horticulture, food processing and packaging, food storage and transportation, food safety, distribution systems, and exports are all central to moving beyond subsistence sustainable agriculture, generating growth, and moving toward prosperity.

Developed and emerging economies have started to identify and support policies and programs to assist Africa in taking a comprehensive approach to agricultural development to break out of poverty. But this will require further rethinking of the agenda to create innovation systems to foster interactions among government, industry, academia, and civil society—all of which are critical actors in agriculture.

The book is guided by the view that innovation is the engine of social and economic development in general and agriculture in particular. The current concerns over rising food prices have compounded concerns about the state and future of African agriculture. This sector has historically lagged behind the rest of the world. Part of the problem lies in the low level of investment in Africa's agricultural research and development. Enhancing African agricultural development will require specific efforts aimed at aligning science and technology strategies with agricultural development efforts. Furthermore, such efforts will need to be pursued as part of Africa's growing interest in regional economic integration through its Regional Economic Communities.

African leaders in recent years have been placing increased emphasis on the role of science and innovation in economic transformation. In July 2014 at the 23rd summit of the African Union, heads of state and government adopted the Science, Technology, and Innovation Strategy for Africa (STISA-2024), which was the result of the High-Level Panel's comprehensive review. The decisions are part of a growing body of guidance on the role of science and innovation in Africa's economic transformation. These decisions underscore the growing importance that African leaders place on science and innovation for development.

However, the translation of these decisions into concrete action remains a key challenge for Africa. This book is guided by the view that one of the main problems facing African countries is aligning national and regional levels of governance with long-term technological considerations. This challenge is emerging at a time when African countries are seeking to deepen economic integration and expand domestic markets. These efforts are likely to affect the way in which agricultural policy is pursued in Africa.

STISA-2024 pays particular attention to the role of science, technology, and innovation in Africa's economic transformation, and it marks the commitment to identifying and building constituencies for fostering science, technology, and innovation in Africa. It focuses on the need to undertake the policy reforms necessary to align the missions and operations of institutions of higher learning with economic development goals in general and the improvement of human welfare in particular.

These decisions represent a clear expression of political will and interest in pursuing specific reforms that would help in making science, technology, and innovation relevant to development. Africa as a whole is making progress, and leaders are starting to take charge. Their capacity to do so would be greatly enhanced by informed advice on international comparative experiences on the subject. This book argues that Africa can feed itself in a generation. There are three opportunities that can help make this vision a reality: advances in science, technology, and engineering—including improvements in infrastructure; better efforts at capacity building through higher technical training; and the emergence of a new crop of entrepreneurial leaders dedicated to the continent's economic improvement.

The book is divided into nine chapters. Chapter 1 examines the critical linkages between agriculture and economic growth. The 2008–09 global economic crisis, rising food prices, and the threat of climate change have reinforced the urgency to find lasting solutions to Africa's agricultural challenges. The entire world needs to find ways to intensify agricultural production while protecting the environment.[9] Africa is composed largely of agricultural economies, with the majority of the population deriving their income from farming. Food security, agricultural development, and economic growth are intertwined. Improving Africa's agricultural performance will require deliberate policy efforts to bring higher technical education, especially in universities, to the service of agriculture and the economy. It is important to focus on how to improve the productivity of agricultural workers, most of whom are women, through technological innovation.

Chapter 2 reviews the implications of advances in science and technology for African agriculture. The Green Revolution played a critical role in helping to overcome chronic food shortages in Latin America and Asia. The Green Revolution was largely a result of the creation of new institutional arrangements aimed at using existing technology to improve agricultural productivity. African countries are faced with enormous technological challenges, but they also have access to a much larger pool of scientific and technical knowledge than was available when the Green Revolution was launched.

It is important to review major advances in science, technology, and engineering and to identify their potential for use in African agriculture. Such exploration should include an examination of local innovations as well as indigenous knowledge. It should cover fields such as information and communication technology, genetics, ecology, and geographical sciences. Understanding the convergence of these and other fields and their implications for African agriculture is important for effective decisionmaking and practical action.

Chapter 3 analyzes the frontiers of agricultural biotechnology, including genetic editing and genetic modification. It examines the benefits of transgenic crops and the challenges of regulating them. Africa must develop its own regulatory system that can analyze each crop on a case-by-case basis and takes into account local context.

Chapter 4 provides a conceptual framework for defining agricultural innovation in a systemic context. The use of emerging technology and indigenous knowledge to promote sustainable agriculture will require adjustments in existing institutions. New approaches will need to be adopted to promote close interactions among government, business, farmers, academia, and civil society. It is important to identify novel agricultural innovation systems of relevance to Africa. This chapter examines the connections between agricultural innovation and wider economic policies. Agriculture is inherently a place-based activity and thus the book outlines strategies that reflect local innovation clusters and other characteristics of local innovation systems. Positioning sustainable agriculture as a knowledge-intensive sector will require fundamental reforms in existing learning institutions, especially universities and research institutes. Specifically, key functions such as research, teaching, extension, and commercialization need to be much more closely integrated.

In Chapter 5 the book outlines the critical linkages between infrastructure and agricultural innovation. Enabling infrastructure (covering public utilities, public works, transportation, and research facilities) is essential for agricultural development. Infrastructure is defined here as facilities, structures, associated equipment, services, and institutional arrangements that facilitate the flow of agricultural goods, services, and ideas. Infrastructure represents a foundational base for applying technical knowledge in sustainable development and relies heavily on civil engineering. The importance of providing an enabling infrastructure for agricultural development cannot be overstated. Modern infrastructure facilities will also need to reflect the growing concern over climate change. In this respect, the chapter will focus on ways to design "smart infrastructure" that takes advantage of advances in the engineering sciences as well as ecologically sound systems design. Unlike other regions of the world, Africa's poor infrastructure represents a unique opportunity to adopt new approaches in the design and implementation of infrastructure facilities.

The role of education in fostering agricultural innovation is the subject of Chapter 6. Some of Africa's most persistent agricultural challenges lie in the educational system. Much of the focus of the educational system is training young people to seek employment in urban areas. Much of the research is carried out in institutions that do not teach, while universities have limited access to research support. But there is an urgency to identify new ways to enhance competence throughout the agricultural value chain, with emphasis on the role of women as farm workers and custodians of the environment. It is important to take a pragmatic approach that emphasizes competence building as a key way to advance social justice. Most of the strategies to strengthen the technical competence of African farmers will entail major reforms in existing universities and research institutions. In this respect, actions need to be considered in the context of agricultural innovation systems.

Chapter 7 presents the importance of entrepreneurship in agricultural innovation. The creation of agricultural enterprises represents one of the most effective ways to stimulate rural development. The chapter will review the efficacy of the policy tools used to promote agricultural enterprises. These include direct financing, matching grants, taxation policies, government or public procurement policies, and rewards to recognize creativity and innovation. It is important to learn from China's Spark Program, which helped to popularize modern technology in rural areas and has spread to more than 90% of the country's counties. Inspired by such examples, Africa should explore ways to create incentives that stimulate entrepreneurship in the agricultural sector. It is important to take into account new tools such as information and communication technologies and the extent to which they can be harnessed to promote entrepreneurship.

Chapter 8 outlines regional approaches for fostering agricultural innovation. African countries are increasingly focusing on promoting regional economic integration as a way to stimulate economic growth and expand local markets. Considerable progress has been made in expanding regional trade through regional bodies such as COMESA, the Southern African Development Community (SADC), and the East African Community (EAC). There are eight other such RECs that have been recognized by the African Union as building blocks for pan-African economic integration.

Chapter 9 concludes the book with suggestions for policy options.

So far regional cooperation in agriculture is in its infancy, and major challenges lie ahead. Africa should intensify efforts to use regional bodies as agents of agricultural innovation through measures such as regional specialization. The continent should explore ways to strengthen the role of the RECs in promoting common regulatory standards.

It is not possible to cover the full range of agricultural activities in one volume. But we hope that the systems approach adopted in the book will help leaders and practitioners to anticipate and accommodate other sources of agricultural innovation.[10]

THE NEW HARVEST

1

THE GROWING ECONOMY

The 2008–2009 global economic crisis, rising food prices, and the threat of climate change have reinforced the urgency to find lasting solutions to Africa's agricultural challenges. Africa is largely an agricultural economy, with the majority of the population deriving their income from farming. Agricultural development is therefore intricately linked to overall economic development in African countries. Most policy interventions have focused on "food security," a term that is used to cover key attributes of food, such as sufficiency, reliability, quality, safety, timeliness, and other aspects of food that are necessary for healthy and thriving populations. This chapter outlines the critical linkages between food security, agricultural development, and economic growth and explains why Africa has lagged behind other regions in agricultural productivity. Improving Africa's agricultural performance will require significant political leadership, investment, and deliberate policy efforts.

The Power of Inspirational Leadership

In a prophetic depiction of the power of inspirational models, Mark Twain famously said, "Few things are harder to put up with than the annoyance of a good example." Africa is starting to see the emergence of African presidents as champions of agricultural transformation. Some of the pioneering advocates included Olusegun Obasanjo (Nigeria), Bingu wa Mutharika (Malawi) and Jakaya Kikwete (Tanzania). Nigeria, under the leadership of former President Goodluck Jonathan, offers another inspirational example of the importance of high-level political support for agricultural transformation. During his tenure he committed his cabinet to making agriculture a primary driver of economic development. He provided leadership by launching the Nigerian Agricultural Transformation Agenda (ATA) in 2011. His vision was that agriculture would be Nigeria's new oil and leading foreign exchange earner. Jonathan's goal for ATA—to add an extra 20 million metric tons of food to domestic food supply by 2015 and to create 3.5 million new jobs—was as bold and ambitious as his overall dream for Nigerian agriculture. To achieve his goal, he hired US-trained and globally respected agricultural economist Akinwumi Adesina, who had decades of experience in African agricultural development. Armed with 25 years of working in some of the foremost international development organizations, including the Rockefeller Foundation, Adesina began the hard work of turning the agriculture sector around.[1]

A fundamental paradigm shift was made to turn agriculture away from a development program. Agriculture is now treated as a business, as a core strategy of Nigeria's strategy to diversify its economy. Adesina, as the dynamic and visionary Minister of Agriculture and the arrowhead of the turnaround of Nigeria's agriculture, has a sharp focus on promoting government-enabled, private sector–led transformation.

ATA is built on five major pillars: (1) efficient input delivery, or the Growth Enhancement Support (GES) scheme; (2) increased value addition, or the value chain approach; (3) building efficient output markets; (4) increased access to finance by farmers; and (5) policy reforms to create incentives for investors in the sector.

In the last three and a half years, ATA has made remarkable progress. A bold reform was implemented to end four decades of corruption in the fertilizer sector, once dominated by direct government procurement and distribution of fertilizers. Within 90 days of being appointed as Minister of Agriculture, Adesina had ended 40 years of corruption in the fertilizer sector. The old system was replaced with a private-sector-driven system. To target farmers effectively, a national database of farmers was developed, and over 14 million farmers have been registered.

Under the GES scheme of the federal government's Agricultural Transformation Agenda, an electronic wallet scheme (e-wallet) was launched for farmers to receive subsidized inputs via an electronic voucher delivered to their cell phones. Nigeria is the first country in the world to launch the e-wallet scheme to deliver subsidized farm inputs at scale. More than 14 million farmers received their subsidized seeds and fertilizers using the e-wallet scheme between 2011 and 2014. During the same period, farmers redeemed a total of 1.37 million tonnes of fertilizer (worth US$1 billion), 102,703 tonnes of improved rice seeds, and 67,991 tonnes of improved maize seeds (worth US$0.3 billion).

The seed sector has undergone a dramatic transformation. Seed production in rice, maize, sorghum, soyabean, and other products rose from about 4,252 tonnes in 2010 to 14,788 tonnes in 2011, 44,487 in 2012, and 149,484 in 2013. The number of private seed companies also grew from 11 in 2011 to about 133 in 2014. Today, about 99 seed companies are participating in the GES program. In addition, global seed industry leaders such as Syngenta, SeedCo West Africa, and DuPont-Pioneer are investing over US$100 million in Nigeria's seed sector.

As a result, in 2011–2014 an additional 7 million tonnes of rice, 12.6 million of maize, 2 million of cassava, 204,000 of sorghum, and 151,000 of soybean were produced. The national food import bill declined from US$ 7.1 billion in 2009 to US$ 4.3 billion in December 2013 and continues to decline. Crop yields also rose; rice yields rose from 1.5 tonnes/ha to over 4 tonnes/ha, and an additional 2 million hectares were put under rice cultivation in the country. Maize yields rose from less than 2 tonnes/ha to 3 tonnes/ha, and as high as 4 tonnes/ha in the maize belt in northern Nigeria.

Nigeria launched a bold effort to become self-sufficient in rice and to reduce its costs to $2.5 billion on rice imports. Between 2012 and 2014, NERICA rice varieties (Faro 44 and Faro 52) were distributed to 6 million rice farmers. Rice-cultivated areas grew by 2 million hectares, while paddy production rose by an additional 7 million hectares.

For the first time, Nigeria launched a bold national policy to support food production in the dry season. Nigerian entrepreneurs responded to the rice revolution by establishing an additional 19 integrated rice mills with parboiling capacity and total combined paddy-milling capacity of 700,000 tonnes. A new rice policy, put in place to encourage local production and milling in place of rice importation, has attracted $2.6 billion of private sector investments. A single investor, Aliko Dangote, Africa's richest man, is investing US$1 billion in the production of 210,000 tonnes of milled rice annually.

A composite flour policy, for import substitution that requires up to 20% inclusion of high quality cassava flour (HQCF) in bread flour, has attracted the participation of wheat millers and industrial and master bakers. Today there is 10% HQCF composite flour from the two largest wheat millers in the country and a universal 2% inclusion of HQCF in bread flour. In addition, 35 industrial and small bakers now produce and market 20% HQCF bread. To ramp up the availability of HQCF, 12 medium-sized HQCF mills are being built to raise HQCF production from less than 30,000 tonnes/ha to 210,000 tonnes/ha per annum, for a universal inclusion of 10% HQCF. To ensure fresh cassava roots at the price and quantity for HQCF production, a total of 29,500 smallholder and 5,300 medium-sized mechanized cassava farms are being established at locations of medium-sized HQCF mills.

There have been interventions via GES in maize, cocoa, oil palm, sorghum, cotton, fisheries and aquaculture, and other value chains, to mention a few. A total of 21,356 tonnes of improved maize seed was distributed to 978,724 registered maize farmers free of charge between 2011 and 2014, and this led to an additional production of 12 million tonnes of maize. For cocoa, 1.5 million cocoa pods, which can produce up of 39 million seedlings, have been distributed to cocoa farmers in 16 states for the rehabilitation of older plantations and to establish new plantations. The objective is to raise cocoa production to 500,000 tonnes per year, up from the 250,000 tonnes per year produced in 2011.

In the oil palm value chain, a total of 9 million sprouted nuts supplied by NIFOR are being distributed to 1,082,831 farmers, comprising 700,331 smallholders and 382,500 medium- to large-estate operators. Under GES, over 5,000 tonnes of seeds of improved varieties, namely Samcot 8, Samcot 9, Samcot 10, and Samcot 11, were distributed free of charge to 174,738 farmers in 2012 and 2013, cultivating an estimated 150,640 hectares of cotton. Working with BOI, 15 ginneries are also being revived to process the cotton harvest from farmers. For sorghum, 515 tonnes of seed was distributed to 102,578 sorghum farmers; 36,711 ha were planted in improved sorghum seeds, and an estimated 55,067 tonnes of sorghum were harvested.

In response to GES support of artisanal fishermen and fish farmers, fish production has risen from 292,105 tonnes to 418,537 tonnes (a 43% increase) for artisanal fishing in inland rivers and lakes, and from 221,128 tonnes to 278,706 tonnes (a 26% increase) for farmed fish. In addition, beef cold chain is being developed in the country. Nigerian entrepreneur Famag-Jal Farms is investing $6 million in a Halal-certified processing plant, with a daily capacity of 300 cows, 890 sheep, and 1,870 goats.

Banks have also increased lending to the agricultural sector, facilitated by NIRSAL, the risk-sharing facility of the Central Bank of Nigeria; a total of US$0.26 billion was lent to fertilizer and seed companies by banks between 2011 and 2014. A US$100 million fund for Agricultural Financing in

Nigeria (FAFIN) was launched for long-term, tailored financing. The fund is capitalized by the Ministry of Agriculture, the German Development Bank (KFW), and Nigeria's Sovereign wealth fund.

The latest release by the National Bureau of Statistics shows that the agricultural sector grew by 9.19% (year-on-year) in the third quarter of 2014, up by 2.7% points from the third quarter of 2013. The agricultural sector grew by 38.53% between the third and fourth quarters of 2014, with crop production being the main driver, with a growth of 43.5%.

The country launched a bold effort to mechanize its agriculture and, in the words of President Jonathan, "put hoes and cutlasses into the museum." To modernize primary production in the country, President Jonathan launched a $340 million farm mechanization policy to establish 1,200 private-sector-driven Agricultural Equipment Hiring Enterprises across the country. Farmers will be provided with mechanized services grant support on mobile phones, which they will use to hire mechanized services. Twenty silo complexes, with a total capacity of 1,025,000 tonnes, are being built nationwide; 10 silos, with capacity of 525,000 tonnes, have been completed. These will be used to develop agricultural commodity exchanges.

To attract private-sector processing companies into locations of high crop production, the concept of Staple Crop Processing Zones (SCPZ), a type of an economic zone, was introduced. Through the development of infrastructure (roads, energy, water, and natural gas), feedstock supply, and provision of fiscal incentives, SCPZs will help to de-risk agro processing for the private sector. In the first phase of 12 SCPZs, there are two in the Southwest (fisheries in Lagos State and cassava in Ogun State); three in the North Central (rice in Niger State, cassava in Kogi State, and fruits in Benue State); two in the Southeast (rice in Anambra and States); three in the Northwest (rice in Kebbi and Sokoto States, tomato and sorghum in Kano State); and two in the Northeast (sorghum in Borno and rice in Taraba).

Development of master plans for the first six SCPZs has been completed and implementation has started for the SCPZs in Kogi and Kano; $1 billion has also been secured from the World Bank and the African Development Bank to execute the master plans. SCPZs are a strategic partnership between the private sector, international development partners, state governments, local communities and ministries, and departments and agencies of the federal government of Nigeria. A highly successful SCPZ event for investors was held at the Abuja World Economic Forum in May 2014.

An Agricultural Resilience Framework has been launched for the creation of climate-smart agriculture. A crop insurance product, Planting with Peace Program, has been introduced to cover farmers' losses due to elemental and weather perils of flood, drought, fire, pests, and diseases. The program will reach 2.5 million farmers with insurance in 2015. The Resilience Framework is also deploying weather stations across the continent to increase their density of coverage; data from these weather stations can be used to supply farmers with more accurate predictions of seasonal weather and short-term rainfall predictions.

These efforts have now made agriculture in Nigeria an exciting sector. The private sector has also committed over $8 billion to existing and planned investments in Nigeria's agriculture, agribusiness, and food industry—with US$4 billion going to expanded fertilizer production. Major agricultural companies such as Cargill, Syngenta, Dupont, Ingredion, and Monsanto have started or are at advanced planning stages of making multimillion US$ investments in Nigeria. Young people are seeing agriculture differently and are going into primary production, processing, and marketing of agricultural products. Adesina has been lauded in Africa and globally for his bold reforms of Nigeria's agriculture, for which he won the prestigious Forbes Africa Person of the Year 2013. Agriculture has indeed become the new oil, as envisaged four years ago by President Jonathan.

Prior to Nigeria's agricultural leadership, Malawi demonstrated remarkable efforts to address the challenges of food security against the rulebook of economic dogma that disparages agricultural subsidies to farmers. Then President Bingu wa Mutharika defied these teachings and put in place a series of policy measures that addressed agricultural development and overall economic development. He serves as an example for other African leaders of how aggressive agricultural investment (16% of government spending) can yield increased production and results.

His leadership should be viewed against a long history of neglect of the agricultural sector in Africa. The impact of structural adjustment policies on Malawi's agriculture was evident from the late 1980s.[2] Mounting evidence showed that growth in the smallholder sector had stagnated, with far-reaching implications for rural welfare. The focus of dominant policies was to subsidize consumers in urban areas.[3] This policy approach prevailed in most African countries and was associated with the continued decline of the agricultural sector.

In 2005, over half of the population in Malawi lived on less than a dollar a day, a quarter of the population lacked sufficient food daily, and a third lacked access to clean water. This started to change when Malawi's wa Mutharika took on food insecurity, a dominant theme in the history of the country.[4] His leadership helped to revitalize the agricultural sector and provides an inspiring lesson for other figures in the region who wish to enable and empower their people to meet their most basic needs.

In 2005, Malawi's agricultural sector employed 78% of the labor force, over half of whom operated below subsistence. Maize is Malawi's principal crop and source of nutrition, but for decades, low rainfall, nutrient-depleted soil, inadequate investment, failed privatization policies, and deficient technology led to low productivity and high prices.[5] The 2005 season yielded just over half of the maize required domestically, leaving five million Malawians in need of food aid.

The president declared food insecurity his personal priority and set out to achieve self-sufficiency and reduce poverty, declaring, "Enough is enough. I am not going to go on my knees to beg for food. Let us grow the food ourselves."[6] The president took charge of the Ministry of Agriculture and Nutrition and initiated a systematic analysis of the problem and potential solutions. After a rigorous assessment, the government designed a program to import improved seeds and fertilizer for distribution to farmers at subsidized prices through coupons.

This ambitious program required considerable financial, political, and public support. The president engaged in debate and consultation with Malawi's parliament, private sector, and civil society, while countering criticism from influential institutions.[7] For example, the International Monetary Fund (IMF) and the US Agency for International Development (USAID) had fundamentally disagreed with the subsidy approach, claiming that it would distort private-sector activities. Other organizations, such as the South Africa–based Regional Hunger and Vulnerability Programme, questioned the ability of the program to benefit resource-poor farmers.[8] On the other hand, the UK Department for International Development (DFID) and the European Union, Norway, Ireland, and later the World Bank supported the program. Additional support came from China, Egypt, and the Grain Traders and Processors Association. The president leveraged this support and several platforms to explain the program and its intended benefits to the public, as well as their role in the system.[9] With support increasing and the ranks of the hungry swelling, the president devoted approximately US$50 million from discretionary funds and some international sources to forge ahead with the program.[10]

The president's strategy attempted to motivate the particularly poor farmers to make a difference not only for their families, but also for their community and their country. Recognizing the benefits of the program, people formally and informally enforced the coupon system to prevent fraud and corruption. The strategy sought to target smallholder farmers, who face the biggest challenges but whose productivity is essential for improving nutrition and livelihoods.[11]

In 2005–2006, the program, coupled with increased rainfall, contributed toward a doubling of maize production, and in 2006–2007, the country recorded its highest surplus ever. Prices fell by half, and Malawi began exporting maize to its food-insecure neighbors. Learning from experience, the government made a number of adjustments and improvements to the program in its first few years, including stepped-up enforcement of coupon distribution, more effective targeting of subsidies, private sector involvement, training for farmers, irrigation investments, and post-harvest support.

President Jonathan's and wa Mutharika's commitment to tackling their nations' most pressing problem and development opportunity is a model for channeling power to challenge the status quo. These are all too rare integrated approaches to studying an issue, developing a solution, and implementing it with full force, despite a hostile international environment, which demonstrate the difference that political will can make.

Innovation and Economic Development

Innovation is at the heart of economic transformation. Joseph Schumpeter's seminal 1911 work, *The Theory of Economic Development*, outlines a general framework for understanding the role of innovation and entrepreneurship in economic development. For Schumpeter, economic development is nonlinear; it arises from endogenous systemic change—not external stimuli—and must take into account more than just economic conditions. At the heart of economic transformation lies creative destruction of the status quo. In a famous example, he explains that adding mail coaches incrementally will not result in the creation of a railway. Instead, he argued, "[the] process of industrial mutation . . . incessantly revolutionizes the economic structure *from within*, incessantly destroying the old one, incessantly creating a new one. This process of Creative Destruction is the essential fact about capitalism."[12] Economic transformation occurs through the creation of new combinations of existing technologies, which includes products as well as processes. Entrepreneurs are best suited to carry out new combinations and disrupt the status quo, and credit-providing institutions should absorb the risk of providing funding to entrepreneurs.

These ideas are particularly salient for emerging countries where agriculture accounts for a large percentage of the economy. Schumpeter was interested in how latecomer countries can catch up. The real benefit of catch-up and leapfrogging lies in path creation, and no sector better embodies the promise of technological leapfrogging than agriculture.

As a sector, agriculture is inherently entrepreneurial. In fact, over the centuries farmers have proven that they are entrepreneurs who are often forced to respond creatively to changes in their conditions. They are most successful when certain foundations of economic transformation are in place. These include infrastructure, training, and credit availability, among others. Of these factors, infrastructure is key. It creates opportunities for entrepreneurs to expand opportunities for new businesses. It also transforms the economic system in a discontinuous way by not only disrupting previous economic practices, but also by expanding opportunities for new economic combinations. Training, or capacity building, is just as important as infrastructure. Appropriate levels of higher and technical education allow entrepreneurs to use existing technologies in new ways to address local problems.

Finally, a systems approach is important in addressing the ecological implications of development. Rather than following traditional conservation efforts that simply seek to minimize human activity, it would be more productive to promote sustainable development—especially in agriculture—through a greater use of innovation.[13]

Linkages Between Agriculture and Economy

Agriculture and economic development are intricately linked. It has been aptly argued that no country has ever sustained rapid economic productivity without first solving the food security challenge.[14] Evidence from industrialized countries, as well as countries that are rapidly developing today, indicates that agriculture stimulated growth in other sectors and supported overall economic well-being. Economic growth originating in agriculture can significantly contribute to reductions in poverty and hunger. Increasing employment and incomes in agriculture stimulates demand for nonagricultural goods and services, boosting nonfarm rural incomes as well.[15] While future trends in developing countries are likely to be affected by the forces of globalization, the overall thesis holds for much of Africa.

Much of our understanding of the linkages between agriculture and economic development has tended to use a linear approach. Under this model, agriculture is seen as a source of input into other sectors of the economy. Resources, skills, and capital are presumed to flow from agriculture to industry. In fact, this model is a central pillar of the "stages of development" that treat agriculture as a transient stage toward industry phases of the economy.[16] This linear view is being replaced by a more sophisticated outlook that recognizes the role of agriculture in fields such as "income growth, food security and poverty alleviation; gender empowerment; and the supply of environmental services."[17] A systems view of economic evolution suggests continuing interactions between agriculture and other sectors of the economy in ways that are mutually reinforcing.[18] Indeed, the relationship between agriculture and economic development is interactive and is associated with uncertainties that defy causal correlation.[19]

The Green Revolution continues to be a subject of considerable debate.[20] However, its impact on agricultural productivity and reductions in consumer prices can hardly be disputed. Much of the debate over the impact of the Green

Revolution ignores the issue of what would have happened to agriculture in developing countries without it. On the whole, without international research in developing countries, yields in major crops would have been higher in industrialized countries by up to 4.8%. This is mainly because lower production in the developing world would have pushed up prices and given industrialized country farmers incentives to boost their production. It is estimated that crop yields in developing countries would have been about 23.5% lower without the Green Revolution and that equilibrium prices would have been 35%–66% higher in 2000. But in reality, prices would have remained constant or would have risen marginally in the absence of international research. This is mainly because real grain prices actually dropped by 40% from 1965 to 2000.[21]

Higher world prices would have led to the expansion of cultivated areas, with dire environmental impacts. Estimates suggest that crop production would have been up to 6.9% higher in industrialized countries and up to 18.6% lower in developing countries. Over the period, developing countries would have had to increase their food imports by nearly 30% to offset the reductions in production. Without international research, caloric intake in developing countries would have dropped by up to 14.4% and the proportion of malnourished children would have increased by nearly 8%. In other words, the Green Revolution helped to raise the health status of up to 42 million preschool children in developing countries.[22]

It is not a surprise that African countries and the international community continue to seek to emulate the Green Revolution or recommend its variants as a way to address current and future challenges.[23] More important, innovation-driven agricultural growth has pervasive economy-wide benefits, as demonstrated through India's Green Revolution. Studies on regional growth linkages have shown strong multiplier effects from agricultural growth to the rural nonfarm economy.[24]

It is for this reason that agricultural stagnation is viewed as a threat to prosperity. Over the last 30 years, agricultural yields and the poverty rate have remained stagnant in sub-Saharan Africa. Prioritizing agricultural development could lead to significant, interconnected benefits, particularly in achieving food security and reducing hunger, increasing incomes and reducing poverty, advancing the human development agenda in health and education, and reversing environmental damage.

In sub-Saharan Africa, the agricultural sector directly contributes to approximately 25% of GDP (or close to half, if the broader sector is included) and 60% of employment. In at least 10 countries, however, the sector accounts for 80%–90% of the workforce.[25] In low-income, resource-poor countries globally, growth in agriculture has been shown to be at least two to four times more effective in reducing the poverty gap than growth in other sectors.[26] In sub-Saharan Africa, it can be 11 times more effective among the extreme poor.[27] As a result, agricultural growth is also highly effective in reducing hunger and malnutrition.[28] Growth in agriculture also stimulates productivity in other sectors such as food processing. Agricultural products comprise about 20% of Africa's exports.[29] Given these figures, it is no surprise that agricultural research and extension services can yield a 35% rate of return, and irrigation projects a 15%–20% return in sub-Saharan Africa.[30]

Even before the global financial and fuel crises hit, hunger was increasing in Africa. In 1990, over 150 million Africans were hungry; as of 2010, the number had increased to nearly 239 million. Starting in 2004, the proportion of undernourished began increasing, reversing several decades of decline, prompting 100 million people to fall into poverty. One-third of people in sub-Saharan Africa are chronically hungry—many of whom are smallholders. High food prices in local markets price out the poorer consumers—forcing them to purchase less food and less nutritious food, as well as diverting spending from education and health and selling their assets. This link between hunger and a weak agricultural sector is self-perpetuating. As a World Bank study has shown, caloric availability has a positive impact on agricultural productivity.[31]

Half of African countries with the highest levels of hunger also have among the highest gender gaps. Agricultural productivity in sub-Saharan Africa could increase significantly if such gaps were reduced in school and in the control of agricultural resources such as land. In addition to this critical gender dynamic, the rural-urban divide is also a key component of the agricultural and economic pictures.

From the 1970s to the 2000s, growth in agricultural gross domestic product (GDP) in Africa has averaged approximately 3%, but there has been significant variation among countries. Growth per capita, a proxy for farm income, was almost zero in the 1970s and negative from the 1980s into the 1990s. Six countries experienced negative per capita growth. As such, productivity has been basically stagnant over 40 years—despite significant growth in other regions, particularly Asia, thanks to the Green Revolution.[32] Different explanations derive from a lack of political prioritization, underinvestment, and ineffective policies. The financial crisis exacerbated this underinvestment, as borrowing externally became more expensive, credit was less accessible, and foreign direct investment declined, although these trends are now starting to show signs of reversal. In recent years, multinational companies are starting to invest in agriculture by sourcing local raw materials, for example, but more must be done to spur new industries and services. This increase in investment has resulted in improved productivity, better incomes, new jobs, and it has helped to open up access to global value chains.

Only 6% of Africa's crop area is irrigated (4% in sub-Saharan Africa), compared to 37% in Asia and 14% in Latin America. Furthermore, more than 40% of the rural population lives in arid or semi-arid conditions, which have the least agricultural potential. Similarly, about 50 million people in sub-Saharan Africa and 200 million people in North Africa and the Middle

East live in areas with absolute water scarcity. Cropland per agricultural population has been decreasing for decades. Soil infertility has occurred due to degradation: nearly 75% of the farmland is affected by excessive extraction of soil nutrients. It is estimated that 60% of the population would benefit from greater irrigation, helping to increase agricultural productivity in sub-Saharan Africa, which is only 56% of the world average.[33]

One way that farmers try to cope with low soil fertility and yields is to clear other land for cultivation. This practice amounts to deforestation, which accounts for up to 30% of greenhouse gas emissions globally. Another factor leading to increased greenhouse gas emissions is limited access to markets: more than 30% of the rural population in sub-Saharan Africa, the Middle East, and North Africa live more than five hours from a market; another 40% live between two to four hours from a market.

Fertilizer use in Africa is less than 10% of the world average of 100 kilograms per hectare. Just five countries (Ethiopia, Kenya, South Africa, Zimbabwe, and Nigeria) account for about two-thirds of the fertilizer applied in Africa. On the average, sub-Saharan African farmers use 13 kilograms of nutrients per hectare of arable and permanent cropland. The rate in the Middle East and North Africa is 71 kilograms. Part of the reason that fertilizer usage is so low is the high cost of imports and transportation; fertilizer in Africa is two to six times the average world price. This results in low usage of improved seed; as of 2000, about 24% of the cereal-growing area used improved varieties, compared to 85% in East Asia and the Pacific. As of 2005, 70% of wheat crop area and 40% of maize crop area used improved seeds, a significant improvement.

Africa's farm demonstrations show significantly higher average yields compared to national yields and show great potential for improvement in maize. For example, Ethiopia's maize field demonstrations yield over five tons per hectare compared to the national average of two tons per hectare for a country plagued by chronic food insecurity. This potential will only be realized as Africans access existing technologies and improve them to suit local needs.

China's inspirational success in modernizing its agriculture and transforming its rural economy over the last 30 years provided the basis for rapid growth and a substantial improvement in prosperity. From 1978 to 2011 China's economy grew at an annual average rate of about 9%. Its agricultural GDP rose by about 4.6% per year, and farmers' incomes grew by 7% annually. Today, just 200 million small-scale farmers, each working an average of 0.6 hectares of land, feed a population of 1.3 billion. In the meantime, China was able to limit population growth at 1.07% per year, using a variety of government policies. Even more remarkable has been the rate of poverty reduction. China's poverty incidence fell from 31% in 1978 to 9.5% in 1990 and then to 2.5% in 2008. Food security has been dramatically enhanced by the growth and diversification of food production, which outstripped population growth. Agriculture's role in reducing poverty has been three times higher than that of other sectors. Agriculture has therefore been the main force in China's poverty reduction and food security.[34]

Lessons from China show that detailed and sustained focus on small-scale farmers by unleashing their potential and meeting their needs can lead to growth and poverty reduction, even when the basic agricultural conditions are unfavorable. But a combination of clear public policies and institutional reforms is needed for this to happen. The policies and reforms need to be adjusted in light of changing circumstances to bolster the rural economy (through infrastructure services, research support, and farmer education), stimulate off-farm employment, and promote rural-urban migration as rural productivity rises and urban economies expand.

With population in check, China's grain production soon outstripped direct consumption, and policy attention shifted to agricultural diversification and improvement of rural livelihoods. The process was driven by a strong, competent, and well-informed developmental state that could set clear medium- and long-term goals and support their implementation.

Despite the historical, geographic, political, social, educational, and cultural differences between China and Africa, there are still many lessons from China's agricultural transformation that can inspire Africa's efforts to turn around decades of low agricultural investment and misguided policies. An African agricultural revolution is within reach, provided the continent can focus on supporting small-scale farmers to help meet national and regional demand for food, rather than relying on expansion of export crops.

While prospects for Africa's global agricultural commodities markets (including cocoa, tea, and coffee) are likely to be brighter than in recent decades, the African food market will grow from US$50 billion in 2010 to US$150 billion by 2030. Currently, food imports are estimated at US$35 billion, up from US$13 billion in the 1990s. Meeting this market with local production will generate the revenue needed to attract additional foreign investment and will help in overall economic diversification. Such a transformation will also help expand overall economic development through linkages with urban areas.

China and the OECD's Development Assistance Committee are helping to disseminate lessons from China's experience among African policymakers and practitioners. But they can go further by contributing to the implementation of agricultural strategies developed by African leaders through the Regional Economic Communities (RECs) and other political bodies. At the very least, they should support efforts to strengthen Africa's capacity for evidence-based policymaking and implementation. This will help to create national and regional capacity for strategic thinking and the implementation of specific agricultural programs.[35]

The State of African Agriculture

Africa has abundant arable land, a strong agricultural work-force, and is seeing increasing investment from both internal and external sources, making the continent ripe for agricultural growth and development.[36] Historically, agriculture has been a low priority for both the public and private sectors. In recent years, however, this has been changing. It is now widely recognized that agriculture accounts for high percentages of both GDP and the workforce in sub-Saharan Africa. The tenth anniversary of the Comprehensive Africa Agriculture Development Programme (CAADP) demonstrates how countries are prioritizing agriculture. Investment in agriculture can generate greater productivity and improve food security.

Despite this growing interest in agriculture, the sector still faces considerable challenges. Productivity has suffered from inconsistent policies and/or ineffective implementation strategies.[37] Through CAADP and initiatives such as Grow Africa, governments and the private sector alike are realizing that transforming African agriculture must address challenges such as "low investment and productivity, poor infrastructure, lack of funding for agricultural research, inadequate use of yield-enhancing technologies, weak linkages between agriculture and other sectors, unfavourable policy and regulatory environments, and climate change."[38]

Sub-Saharan Africa faces other challenges as well. Improving agricultural productivity in sub-Saharan Africa will differ remarkably from the path pursued in Asia, for example, which is highly irrigated. Sub-Saharan Africa depends heavily on rainfall (96% of farmers), and it is also prone to weather shocks and diverse growing conditions.[39] Markets are scattered and difficult to access for many farmers. Much of rural Africa is still without passable roads, translating to high transportation costs and trade barriers. In fact, only 40% of rural Africans live within 2 kilometers of an all-weather road, and transportation costs are about 50% higher for landlocked countries.

This inaccessibility greatly hinders agricultural productivity.[40] Although long distances and low population densities make trade, infrastructure, and service provision difficult, they also offer opportunities for expansion.

Human technical capacity at both the farm level and country level is limited, constraining agricultural growth, poverty reduction, and food security throughout the continent.[41] According to the Food and Agriculture Organization (FAO) of the United Nations, education is necessary for a productive agricultural sector. Evidence shows that "an additional year of schooling for the whole population can raise . . . productivity by 3.2%."[42] CAADP policies have faced implementation hurdles due to a lack of human capital. At its core, agriculture is knowledge-based and entrepreneurial. Existing educational opportunities do not tap into this mindset, nor does the educational system in Africa specifically address the challenges faced by the sector.

Sub-Saharan Africa still ranks the lowest in the world in terms of yield-enhancing practices and techniques, even compared to other developing regions. These practices include mechanization, use of agro-chemicals, improved seed, precision farming techniques, and increased use of irrigation.[43]

Mechanization is very low, with approximately 13 tractors per 100 square kilometers of arable land, versus 200 tractors per 100 square kilometers globally.[44] Similarly, there are only two tractors per 1,000 farmers in sub-Saharan Africa versus 883 per 1,000 in the United Kingdom.[45,46] As mentioned above, only 4% of sub-Saharan Africa's cropland is irrigated, compared with the world average of 18.4%, and the use of chemical inputs is minimal, at nine kilograms per hectare compared with the world average of 100 kilograms per hectare. Furthermore, when mechanization and inputs are applied, they are typically only applied in more developed areas with large-scale farms or where market demand is high.

The decades of neglect of African agriculture led to low output. Land productivity in Africa is estimated at 42% and

50% of that of Asia and Latin America, respectively, and 54% of the world's output. This is generally attributed to these regions' use of yield-enhancing practices and techniques.[47]

The potential for improved productivity in Africa is huge, however. Estimates vary slightly, but only about 29% (183 million ha) of the total arable land available in sub-Saharan Africa (635 million ha) is currently under cultivation.[48] Furthermore, between 1970 and 2010, cultivation increased from 132 million to 184 million hectares. Irrigation also increased from 2.4% to 5.3%. Compared to other developing regions, however, this is still extremely low. In fact, because irrigation is so low, only 3.8% of Africa's surface and groundwater is harnessed. As a percentage of its renewable resources, sub-Saharan Africa's water use is just 1%, whereas North Africa's is 219%.[49] North Africa suffers from overuse and unsustainable irrigation practices, but sub-Saharan Africa could greatly benefit from better irrigation practices using new technologies. Productivity could also be improved by linking agriculture with other sectors and building the agricultural value chain. However, area expansion per se should not be a priority in view of increased environmental degradation on the continent. Currently, Africa accounts for 27% of the world's land degradation and has 500 million hectares of moderately or severely degraded land. Soil degradation affects 65% of cropland and 30% of pastureland, and is associated with low land productivity. It is caused by loss of vegetation and land exploitation, especially overgrazing and shifting cultivation.[50]

In spite of these challenges, however, and despite upheavals in the global financial system, Africa continues to register remarkable growth prospects. While African economies continue to face serious challenges, such as poverty, diseases, and high rates of infant mortality, Africa has grown dramatically since 2000. Africa's collective per capita GDP is now at US$953, and almost half of African countries have achieved middle-income status (26 out of 54, up from just 13 in 2006). Another 10 countries could achieve that status by 2025.[51] If current trends continue, by 2015 Africa could achieve a collective GDP of US$2.8 trillion, doubling what it was in 2008.[52] Furthermore, sub-Saharan Africa is the most rapidly growing economic region in the world today. Collectively, its GDP grew at or around 5% per year from 2000 to 2012, reaching almost 6% in 2013. These figures are higher when South Africa is excluded. The poverty rate dropped from 51% to 39%. Six of the world's 10 fastest-growing countries are in sub-Saharan Africa: Angola, Nigeria, Ethiopia, Chad, Mozambique, and Rwanda. Several others were at or above 7%. If this growth trend continues, economies and incomes will double over the next two decades. This is a stark contrast to the 1980–2000 period, when incomes contracted by 20%.[53] With more Africans achieving middle-income status, internal demand is also fueling growth. Major growth sectors include agriculture (e.g., in Ethiopia, where agriculture comprised 42% of its GDP, and Rwanda), services (in Burkina Faso, Tanzania, and Uganda), telecommunications, banking, retail, construction, tourism, and foreign investment. The priority now is on sustaining this growth through widespread economic transformation. Indeed, this is the focus of the African Union (via its "Agenda 2063") and the African Development Bank.

Although this section has depicted the broad trends in African agriculture, "the diversity across sub-Saharan African countries and across regions within countries is huge in terms of size, agricultural potential, transport links, reliance on natural resources, and state capacity."[54] The policy agenda will have to be carefully tailored to country-specific circumstances.

Trends in Agricultural Renewal

Future trends in African agriculture are going to be greatly influenced by developments in the global economy as well as emerging trends in Africa itself. The agricultural sector is the obvious starting point for sustainable economic transformation in sub-Saharan Africa. Agriculture is a primary driver of industrialization and greater economic development. It can increase rural incomes and exports, making it easier to import machinery and other inputs. It can supply the raw materials to support local agricultural-processing industries. As productivity improves, labor can shift from rural to urban areas. Similarly, it can boost the supply of food to these growing urban areas. Finally, a more vibrant agricultural sector can help to expand the markets for inputs and services for the non-agricultural sectors.[55] It is clear why the African Union is prioritizing agriculture, taking a holistic approach and focusing on improving food security and nutrition, promoting agro-processing and the business climate, introducing aquaculture and livestock programs, and improving mechanization.[56]

Advancements in science have demonstrated the important role that niche crops can play in improving human health in sub-Saharan Africa specifically. Achieving food security depends not only on increasing production but also on improving nutrition. Increasing the production of niche crops—also known as ancient grains, orphan crops, lost crops, famine crops, local crops, neglected crops, or wild foods—is one way to achieve this. Technological advancements in agricultural biotechnology and advances in fields such as plant genomics allow for the enhancement of existing crops and the ability to breed new ones that meet higher nutritional standards. Furthermore, many communities rely on niche crops, so increasing their production would also improve nutrition in food-insecure areas.[57]

Agro-processing also has huge potential for changing the sector. Africa already has a comparative advantage in both traditional exports such as coffee, cocoa, and cotton, as well as some nontraditional but high-demand exports such as pineapple and other fruits. Establishing agro-processing businesses locally would add value, create jobs, improve local markets, and earn foreign exchange. The right policies and investments can help jump-start a processing base. Agro-processing also offers the opportunity to scale, which is crucial to economic transformation. Improving the supply chain by connecting farms to processing facilities would also help leapfrog into other export sectors. Finally, establishing agro-processing facilities would help curb imports such as soybean cakes and other livestock inputs. With the rising urban middle class throughout Africa, the demand for poultry and meat will continue to grow. This represents a major source of transformation for African agricultural economies.[58]

It is important to note that agricultural renewal is not all about crops. Aquaculture and livestock are equally significant. Aquaculture is a key source of protein for domestic consumption, as well as an important export. Global demand for fish is rising, and it is estimated that two-thirds of this demand will be met from farmed fish.[59] Much of the demand—70%—will originate in Asia. Many countries are turning to aquaculture. In several coastal countries in sub-Saharan Africa, such as Senegal and Ghana, fish comprise about half of the protein intake and is an important export, but aquaculture remains an underutilized resource overall. Yet aquaculture is an important source of employment, protein, and economic opportunities, especially for small-scale fishing communities. In Ghana, Kenya, Namibia, Nigeria, and Uganda, aquaculture has grown from 55,690 tonnes in 2009 to 600,000 tonnes in 2010, where specific programs have supported the growth of the industry. In other countries such as Burkina Faso, Mali, and Madagascar, farmers are integrating aquaculture with their rice-farming operations.[60] With the right incentives and policies, aquaculture could become an important source of both local consumption and exports throughout sub-Saharan Africa.

Finally, mechanization, too, is a source of both agricultural renewal and climate change mitigation and brings with it a host of benefits. By now, African countries have accepted that mechanization is a key factor in intensifying production and modernizing the sector. A 2014 report notes that sub-Saharan Africa has the lowest mechanization rate in the world, with "motorized equipment [contributing] only about 10% of farm energy" in sub-Saharan Africa, versus 50% in other areas.[61] Old challenges persist: maintenance and repair capabilities are limited; spare parts are hard to acquire; farmers need training in mechanization; and the demand for mechanization outstrips the supply.[62] Governments, however, understand that this is an area where improvements will yield direct and immediate benefits, and they have started investing in mechanization. One benefit is that it improves labor productivity and quality of life for farmers. It improves crop productivity by allowing farmers to plan exactly when to till, plant, weed, and apply fertilizer, based on changing conditions, without relying on labor or other resources. It can help in more appropriate applications (e.g., split application) of organic and nonorganic fertilizer, resulting in healthier soils. It is also a key component of precision agriculture such as direct seeding, which helps to limit tillage and carbon dioxide emissions. Finally, countries such as Uganda, Tanzania, and Zambia are realizing the benefits of technical education and are encouraging greater enrollment in standard and nonstandard programs.

As more technical support becomes available in rural areas, tractors and other equipment become easier to maintain, and new entrepreneurial activities unfold. One example of this comes from Ghana, where a vibrant parallel market has developed alongside government programs promoting mechanization. Smallholder and private-sector farmers who do own tractors are beginning to lease the tractors and their services. Data from Ghana are revealing. The majority of tractors are used, and most are acquired via private funds—only 11% are acquired through government programs. Farmers tend to buy tractors after having hired mechanization services for approximately 10 years; they report expansion as a motivating factor in the purchase. Some of these farmers buy tractors even if they do not have large areas of land under cultivation, with the intent of hiring out their machines. Tractor owners typically plow 20 hectares of their own land, and then lease out services to 60 other farmers, averaging another 160 hectares. Tractors offer an additional benefit: they help with maize shelling as well. This example shows that a vibrant market is likely to develop alongside increased mechanization. Farmers are willing to invest in tractors on their own, which in turn can help to scale up production. Governments can supplement this with research and training in both operational and maintenance skills.[63]

As agricultural growth has huge potential for companies across the value chain, overcoming various barriers to productivity (such as a lack of advanced seeds, inadequate infrastructure, trade barriers, unclear land rights, lack of technical assistance, and finance for farmers) is key to increasing the agricultural output from US$280 billion to a projected US$880 billion by 2030.[64] Focusing on nutrition, agroprocessing, livestock and fisheries, and mechanization is a good way to increase output and improve the lives of farmers. The picture is therefore promising though uncertain. Growth in Africa's collective agricultural GDP has been higher than 4% since 2003. The recent successes in Ethiopia, Nigeria, and Rwanda demonstrate that the link between farm productivity and income growth for the poor indeed operates in Africa. Agriculture, aquaculture, agro-processing, and mechanization represent just four ways that the agricultural sector in sub-Saharan Africa is experiencing renewal. These recent gains in agriculture can be attributed to a better policy environment, increased usage of technology, and higher commodity prices. There are numerous cases that illustrate the ingenious and innovative ways that Africans are overcoming the constraints identified above to strengthen their agricultural productivity and livelihoods.

Enabling Policy Environment

The African Union is not alone in prioritizing agriculture. The African Development Bank and the World Economic Forum are also prioritizing agriculture from a public-private partnership perspective.

Countries that have relaxed constraints (such as over-taxation of the agricultural sector) have been able to increase agricultural productivity. For example, a 10% increase in coffee prices in Uganda has helped reduce the number of people living in poverty by 6%.

The initiation of the African-led CAADP by the African Union's New Partnership for Africa's Development (NEPAD) constitutes a significant demonstration of commitment and leadership. CAADP is uniquely African, and it has encouraged high-level leaders to sign "compacts" in which they pledged to invest at least 10% of their national budgets in the agricultural sector with the goal of raising productivity and growth by 6%. Since 2003, CAADP has been working with the RECs and through national roundtables to promote sharing, learning, and coordination to advance agriculture-led development. CAADP focuses on sustainable land management, rural infrastructure and market access, food supply and hunger, and agricultural research and technology. As of March 2014, 33 countries had signed CAADP compacts, and 25 had developed comprehensive financing plans.[65] The compacts are products of national roundtables at which priorities are set and road maps for implementation are developed. The compacts are signed by all the key partners.

Leading countries include Burkina Faso, which has achieved 13% expenditure rate, accounting for 14% of budget; this was spent on irrigation and farmer field schools. Tanzania focused on production and commercialization: it supported technology transfer; encouraged public-private partnerships that opened the door for ambitious entrepreneurs such as Yara; and invested in infrastructure, which has opened doors in its southern islands. Malawi has spent an average of 10% since 2003. It focuses on self-sufficiency but also on addressing post-harvest losses via grain banks, as well as on water management.

In eastern and southern Africa, the Common Market for Eastern and Southern Africa (COMESA) coordinates the

CAADP planning and implementation processes at country and regional levels. In doing so, it also collaborates with regional policy networks, such as the Food, Agriculture and Natural Resources Policy Analysis Network, and subregional knowledge systems, such as the Regional Strategic Analysis and Knowledge Support Systems, and it utilizes analytical capacity provided through various universities in the region, supported by Michigan State University.

In close coordination with national CAADP processes, a regional CAADP compact is being developed. Its aim is to design a Regional Investment Program on Agriculture that will focus on developing key regional value chains and integrating value chain development into corridor development programs. At a national level, the priority programs developed include those in the area of research and dissemination of productivity-enhancing technologies to promote knowledge-based agricultural practices, applying the innovation systems approach to develop and strengthen linkages between generators, users, and intermediaries of technological knowledge.

Building on the momentum of CAADP, in 2011 the African Union, NEPAD, and the World Economic Forum (WEF) jointly launched Grow Africa, an African-owned partnership platform designed to leverage and encourage private-sector investment in African agriculture. Grow Africa is a public-private partnership platform that aims to make agriculture the engine of growth in Africa by supporting the goals of CAADP; for example, it helps partner countries generate private-sector investment in order to achieve 6% agricultural growth. It builds on the WEF's New Vision for Agriculture Initiative. Spring 2012 saw the emergence of the G-8's New Alliance for Food Security and Nutrition (New Alliance). Through Grow Africa, the New Alliance seeks to link both foreign and African private-sector companies to specific investment needs and opportunities identified by 10 Grow Africa partners—Burkina Faso, Côte d'Ivoire, Ethiopia, Ghana, Kenya, Malawi, Mozambique, Nigeria, Rwanda, and Tanzania—in their CAADP plans.[66] Grow Africa's goal is to generate specific investment commitments by companies in the private sector, while government leaders pledged to improve the policy and business environments. Sixty-two companies initially agreed to participate, resulting in "over $3.5 billion of planned investment," as noted in their letters of intent, which are designed to "create productive dialogue with governments and other partners" and are later used as a benchmark against which to measure progress.[67] As of 2014, commitments had reached $7.2 billion, with $970 million invested and the rest planned for the next three to five years. Much of this investment stems from African companies. Those companies that have successfully invested have reached approximately 2.6 million smallholder farmers, have generated 33,000 jobs, and are already planning to scale up their businesses.[68]

To combat claims that it facilities land grabbing,[69] Grow Africa specifically calls on the private sector to invest in agriculture but also to address the plight of smallholder farmers by integrating them into the market.[70] For example, the ONE Campaign found that most of the companies plan to source crops from smallholders;[71] many are also planning to incorporate some element of training for smallholders into their business plans. Given the general lack of infrastructure in rural areas, a guarantee of payment at the time of harvest—for example, through an outgrower scheme or a warehouse receipt payment system, among others—is highly desirable where market access is limited and many smallholders in one region grow similar crops.

The Land Debate

In a continent that has 60% of the world's arable land, where initiatives such as CAADP and Grow Africa are flourishing and a more enabling policy environment has emerged, interest in investing in African agriculture has risen dramatically while generating debates about land grabbing. Criticism of

these new investments ranges from a new form of colonial-
ism at best to pure land grabbing at worst. Although there is
legitimate concern about these land deals and their long-run
developmental prospects, there is also a strong argument for
attracting responsible agricultural foreign direct investment
and regulating it as such. If done correctly, these types of
investments have the potential to catalyze agricultural development by increasing the productivity of undercultivated
areas, investing in rural infrastructure (including irrigation,
roads, and much-needed storage facilities), and increasing
market access for smallholder farmers. In fact, large-scale
land deals are not new: there are examples of responsible (as
well as irresponsible) agribusiness investments dating back
to the 1960s. These examples can be drawn on as an example of how to properly manage large-scale agricultural deals
today.

Africa is committed to promoting agricultural development
as part of larger strategies to stimulate economic transformation, and attracting foreign investment can help governments
to meet those goals. The FAO's "Voluntary Guidelines on the
Responsible Governance of Tenure of Land, Fisheries, and Forests in the Context of National Food Security" and the African
Union's "Declaration on Land Issues and Challenges in Africa"
are two starting points that identify best practices that governments can follow when managing land and water resources.

There are many ways in which local governments can work
to attract investment while making sure smallholder farmers
also benefit. For example, in 2011 Ethiopia created its own Agricultural Transformation Agency, following Brazil's Agricultural Research Corporation. The government also decided to
lease three million hectares of arable land to private investors
over the next four years. Of Ethiopia's 74 million hectares of
total arable land, only 15 million hectares are currently cultivated. The three million hectares the government hopes to
lease are essentially a modest step in Ethiopia's effort to foster
economic transformation.

Countries that are attracting foreign investment in agriculture are also starting to focus more seriously on reforming their land tenure systems. Recognizing customary land rights and any claims by statutory land laws is a good first step. When national governments take charge of the land policy, they encourage dialogue and consultation between stakeholders and local populations. In Ghana, for example, the government has recently decreed that any lease over 50 hectares (whether the lease holder is foreign or national) must be approved by the national government. In Kenya, the government has prevented investors from leasing large tracts of land without first conducting exhaustive consultations with the local people; Kenya is also considering the need to harmonize land laws with the National Land Policy and the Constitution.

For less than a few dollars, land-use certificates can be implemented to reduce encroachment and improve soil conservation. For example, Ethiopia's system for community-driven land certification has been one effective way to improve land practices and a potential step toward the much broader reform of land policy that is needed in many African countries.

Here is how it works: communities learn about the certification process and then elect land-use committees. These voluntary committees settle conflicts and designate unassigned plots through a survey, setting up a system for inheritable rights. In a nationwide survey, approximately 80% felt that this certification process effectively fulfilled those tasks, as well as encouraged their personal investment in conservation and women's access to resources. The certificates themselves cost US$1 per plot but increase to less than US$3 with mapping and updating using global position system (GPS). Between 2003 and 2005, six million households were issued certificates, demonstrating the scalability.[49] Documenting land rights in this participatory and locally owned way can serve as a model for governments that are ready to take on meaningful reform.

There are many safeguards that governments can implement in order to guard against exploitative land deals. For example, anti-speculative requirements could include the following: (1) establishing a community development fund, where investors must make deposits up front and on a regular basis even before cultivation starts, as well as requiring any investor to come up with a significant down payment, will help to attract serious investors only; (2) creating a tax provision will prevent land from lying dormant in anticipation of future gain; (3) holding auctions for available land can facilitate bidding and increase transparency (as in Peru); and (4) requiring compulsory delivery and/or a concession cancellation would force investors to take possession of the land they acquire within a specified time frame (e.g., Ethiopia can cancel a concession if it is not implemented within 6 months; the Democratic Republic of Congo goes a step further by requiring that the land not only be occupied within 6 months, but it must be put to productive use within 18 months; and in Mozambique an investor has 120 days after the project is authorized to begin implementation).

Governments can also implement other mutually beneficial requirements. These include food production–sharing, outgrower schemes, joint equity with local communities, provisions for infrastructure development, protocols for access to water, public-private partnerships, social agreements with affected communities, direct and indirect job requirements, taxes to both local and national government bodies, extension services, local consultation before a deal is signed, social and environmental impact assessments, monitoring and evaluation components, and prohibition of lease transfer.

Many sub-Saharan African countries are already following some of these provisions when engaging in land deals. Large-scale land acquisitions, when managed properly, are one method of encouraging agricultural development. They should not be automatically ruled out, nor should they be blindly promoted. Every deal requires careful analysis, as each one is situated in a unique local context and must be evaluated accordingly.

Regional Imperatives

The facilitation of regional cooperation is emerging as a basis for diversifying economic activities in general, and leveraging international partnerships in particular. Many of Africa's individual states are no longer viable economic entities; their future lies in creating trading partnerships with neighboring countries.

Many African countries are either relatively small or landlocked, thereby lacking the financial resources needed to invest in major infrastructure projects. Their future economic prospects depend on being part of larger regional markets. Increased regional trade in agricultural products can help them stimulate rural development and enhance their technological competence through specialization. Existing RECs offer them the opportunity to benefit from rationalized agricultural activities. They can also benefit from increased harmonization of regional standards and sanitary measures.[72]

African countries have adopted numerous regional cooperation and integration arrangements, many of which are purely ornamental. The roles of bigger markets in stimulating technological innovation, fostering economies of scale arising from infrastructure investments, and the diffusion of technical skills into the wider economy are some of the key gains that Africa hopes to derive from economic integration. In effect, science and innovation are central elements of the integration agenda and should be made more explicit.

The continent has more than 20 regional agreements that seek to promote cooperation and economic integration at subregional and continental levels. They range from limited cooperation among neighboring states in narrow political and economic areas to the ambitious creation of an African common market. They focus on improving efficiency, expanding the regional market, and supporting the continent's integration into the global economy. Many of them are motivated by factors such as the small size of the national economy, a landlocked position, and poor infrastructure. Of all Africa's regional agreements, the African Union formally recognizes eight RECs. These RECs represent a new economic governance system for Africa and should be strengthened.

The Common Market for Eastern and Southern Africa, in particular, illustrates the importance of regional integration in Africa's economic development and food security. The 19-member free trade area was launched in 2000 and at 420 million people accounts for nearly half of Africa's population. It has a combined GDP of US$475 billion and is the largest and most vibrant free trade area in Africa, with intra-COMESA trade estimated at US$17 billion in 2011. COMESA aims to improve economic integration and business growth by standardizing customs procedures, reducing tariffs, encouraging investments, and improving infrastructure. COMESA launched its customs union on June 7, 2009, in Victoria Falls Town and has initiated work on a Common Investment Area to facilitate cross-border and foreign direct investment. COMESA plans to launch its common market in 2015, and in this regard it already has a program for liberalization of trade in services. The program prioritizes liberalization of infrastructure services, namely, communication, transport, and financial services. Other subsectors will be progressively liberalized.

The strength of the RECs lies in their diversity. Their objectives range from cooperation among neighboring states in narrow political and economic areas to the ambitious creation of political federations. Many of them are motivated by factors such as the small size of the national economy, a landlocked position, or poor infrastructure. Those working on security, for example, can learn from the Economic Community of West African States (ECOWAS), which has extensive experience dealing with conflict in Côte d'Ivoire, Liberia, and Sierra Leone.[73]

Other RECs have more ambitious plans. The East African Community (EAC), for example, has developed a road map that includes the use of a common currency and creation of single federal state. In July 2010 the EAC launched its Common Market by breaking barriers and allowing the free movement of goods, labor, services, and capital among its member states. The EAC Common Market has a combined GDP of US$76 billion. Through a process that began with the establishment of the EAC Customs Union, the Common Market is the second step in a four-phase road map to make the EAC the strongest economic, social, cultural, and political partnership in Africa. EAC's economic influence extends to neighboring countries such as Sudan, Democratic Republic of the Congo, and Somalia. The Common Market will eliminate all tariff and nontariff barriers in the region and will set up a common external tax code on foreign goods. It will also enhance macroeconomic policy coordination and harmonization as well as the standardization of trade practices. It is estimated that East Africa's GDP will grow 6.4% in 2011, making it the fastest growing region in Africa.[74]

The region has already identified agriculture as one of its strategic areas. In 2006 the EAC developed an Agriculture and Rural Development Policy that provides a framework for improving rural life over the next 25 years by increasing the productivity output of food and raw materials, improving food security, and providing an enabling environment for regional and international trade. It also covers the provision of social services such as education, health, and water, and the development of support infrastructure, power, and communications. The overall vision of the EAC is to attain a "well developed agricultural sector for sustainable economic growth and equitable development."[75] Its mission is to "support, promote and facilitate the development, production and marketing of agricultural produce and products to ensure food security, poverty eradication and sustainable economic development."[76] Such institutions, though nascent, represent

major innovations in Africa's economic and political governance and deserve the fullest support of the international
community.

The Growing Impact of Climate Change

All of the strategies described above must take into account
the emerging concerns regarding the impact that climate
change will have on African agriculture—and vice versa. Climate change is affecting growing seasons, yields of staple food
crops, and farming systems throughout the world, but many
developing countries, including those in sub-Saharan Africa,
will be particularly hard hit. Rising temperatures and changing rainfall patterns are already forcing farmers to deal with
myriad problems such as floods, drought, and other weather
events. But even small changes—such as an increase in temper-
ature by 2°C—can "reduce total crop production by 10% . . . by
2050, and increase the undernourished population by at least
25%."[77] Models show that Kenya, South Sudan, and Uganda
will experience drying; Southern Africa will have to adapt to a
delay in summer rains as well as drought; and throughout the
subcontinent yields of staple crops such as maize, sorghum,
and millet are projected to decrease. Pastoralists, too, will have
to find ways to adapt in the face of receding water sources and
a loss of forage areas. Farmers who rely heavily on rainfall are
at the greatest risk.

Farmers in sub-Saharan Africa are finding ways to adapt
and mitigate the effects of climate change. In Burkina Faso
and Niger, for example, farmers use a combination of techniques to improve cereal yields. These techniques include
water-harvesting tools such as stone lines and improved
planting pits, as well as the precise application of fertilizer to
the new plant early in the growing stage. In Malawi, farmers
plant shade trees that improve soil health as one strategy to
increase maize yields. In West Africa, farmers apply a combination of fertilizer, compost, mulch, and manure to increase

soil fertility. This strategy boosted yields from 33% to 58% over four years.[78] There are also longer-term adaptation strategies that focus on "systems adaptation," or making different choices about what to plant and when, as well as livestock changes.[79]

However, much more needs to be done in terms of both mitigation and adaptation. One crucial element that so many farmers lack is risk insurance. Index-based weather, crop, and livestock insurance offers one solution, and pilot projects have sprung up in Ethiopia, Kenya, and Senegal. These climate-smart solutions will be even more effective at helping smallholder farmers make informed decisions as greater investments and innovations are made in information and communications systems.

Conclusion

This chapter has examined the critical linkages between agriculture and economic development in Africa. It opened with a discussion of the importance of inspirational leadership in effecting change. This is particularly important because much of the large body of scientific and technical knowledge needed to promote agricultural innovation in Africa is available. It is widely acknowledged that institutions play an important role in shaping the pace and direction of technological innovation in particular and economic development in general. Much has been written on the need to ensure that the right democratic institutions are in place as prerequisites for agricultural growth.[80] But emerging evidence supports the importance of entrepreneurial leadership in promoting agricultural innovation as a matter of urgency, rather than waiting until the requisite institutions are in place.[81] This view reinforces the important role that entrepreneurial leadership plays in fostering the co-evolution between technology and institutions.

Fundamentally, "it would seem that one can understand the role of institutions and institutional change in economic growth only if one comes to see how these variables are connected to technological change."[82] This is not to argue that institutions and policies do not matter. To the contrary, they do and should be the focus of leadership. But the focus should also be on innovation. The essence of entrepreneurial leadership points to the urgency of viewing institutions and economic growth as interactive and co-evolutionary. The rest of this book will examine these issues in detail.

2

ADVANCES IN SCIENCE, TECHNOLOGY, AND ENGINEERING

The Green Revolution played a critical role in helping to overcome chronic food shortages in Latin America and Asia. The Green Revolution was largely a result of the creation of new institutional arrangements aimed at using existing technology to improve agricultural productivity. African countries are faced with enormous technological challenges. But they also have access to a much larger pool of scientific and technical knowledge than was available when the Green Revolution was launched in the 1950s. Current technological advances in the agricultural sector have the potential to contribute to both food security and food safety. Technology will also be important to in mitigating climate change by reducing the agricultural sector's dependence on fossil fuels and its contribution to global greenhouse gas emissions. Use of biopolymers to encapsulate fertilizer, contributing to slow release in the soil, is one example. Use of this technology can optimize the input of fertilizer and reduce greenhouse gas emission from leakage in the soil and from fertilizer production. The aim of this chapter is to review major advances in science, technology, and engineering and to identify their potential for use in African agriculture. This exploration will also include an examination of local innovations as well as indigenous knowledge. It will cover fields such as information and communications technology, ecology, and geographical sciences. It will emphasize the

convergence of these and other fields and their implications
for African agriculture.

Innovation and Latecomer Advantages

Since the Industrial Revolution there has been an exponential
growth in knowledge. Technological advances through science and engineering have led to new technological discoveries in a speed unthinkable some decades ago. Until recently,
growth was thought to be a linear process. In today's environment, change happens quickly and growth is "global and
exponential." Peter Diamandis and Steven Kotler explain it
best in their book *Abundance*: "If I take thirty linear steps (calling one step a meter) from the front door of my Santa Monica
home, I end up thirty meters away. However if I take exponential steps (one, two, four, eight, sixteen, thirty-two, and so on),
I end up a billion meters away, or, effectively lapping the globe
twenty-six times."[1] This is the direction in which technological development is moving. Growth in technologies has been
characterized as a dynamic process in which new technological discoveries advance our knowledge to create yet more new
technologies and innovations. Recombination of different
pieces of existing technologies creates again new technologies.
In this way, our knowledge grows exponentially. This is referred to as the evolution of technologies.[2] While it is easy to be
intimidated by the pace of technological growth, it could also
be a means to achieve a society of abundance, where all people
have access basic needs, education, freedom, and good health.[3]
Exponential growth in knowledge is also one of the main criteria for what some have called the new machine age, a time
when digital technologies can create self-driving cars and contribute to solve some of the world's most pressing challenges.[4]

African countries can utilize the large aggregation of knowledge and know-how that has been amassed globally in their
efforts to improve their access to and use of the most cutting-
edge technology. While Africa is currently lagging in the

utilization and accumulation of technology, its countries have the ability not only to catch up to industrial leaders but also to attain their own level of research growth.

Advocates of scientific and technical research in developing countries have found champions in the innovation platforms of nanotechnology, biotechnology, information and communication technology (ICT), and geographic information systems (GIS). Through these four platform technologies, Africa has the opportunity to promote its agenda concurrent with advances made in the industrialized world. This opportunity is superior to the traditional catching-up model, which has led to slower development and has kept African countries from reaching their full potential. These technologies are able to enhance technological advances and scientific research while expanding storage, collection, and transmission of global knowledge. This chapter explores the potential of ICT, GIS, nanotechnology, and biotechnology in Africa's agricultural sector and provides examples of where these platform technologies have already created an impact.

Contemporary history informs us that the main explanation for the success of the industrialized countries lies in their ability to learn how to improve performance in a variety of fields—including institutional development, technological adaptation, trade, organization, and the use of natural resources. In other words, the key to success is putting a premium on learning and improving problem-solving skills.[5]

Every generation receives a legacy of knowledge from its predecessors that it can harness for its own advantage. One of the most critical aspects of a learner's strategy is using that legacy. Each new generation blends the new and the old and thereby charts its own development path within a broad technological trajectory, making debates about potential conflicts between innovation and tradition irrelevant.[6]

At least three key factors contributed to the rapid economic transformation of emerging economies. First, these countries invested heavily in basic infrastructure, including roads, schools, water, sanitation, irrigation, clinics, telecommunications, and energy.[7] The investments served as a foundation for technological learning. Second, they nurtured the development of small and medium-sized enterprises (SMEs).[8] Building these enterprises requires developing local operational, repair, and maintenance expertise, and a pool of local technicians. The ability to create an environment where SMEs can scale up and grow into a sustainable industry is also crucial.[9] Third, government supported, funded, and nurtured higher education institutions, as well as academies of engineering and technological sciences, professional engineering and technological associations, and industrial and trade associations.[10]

The emphasis on knowledge should be guided by the view that economic transformation is a process of continuous improvement of productive activities, advanced through business enterprises. In other words, government policy should focus on continuous improvement aimed at enhancing performance, starting with critical fields such as agriculture for local consumption and extending to international trade.

This improvement indicates a society's capacity to adapt to change through learning. It is through continuous improvement that nations transform their economies and achieve higher levels of performance. Using this framework, with government functioning as a facilitator for social learning, business enterprises will become the locus of learning, and knowledge will be the currency of change.[11] Most African countries already have in place the key institutional components needed to make the transition to being a player in the knowledge economy. The emphasis should therefore be on realigning the existing structures and creating new ones where they do not exist.

The challenge is in building the international partnerships necessary to align government policy with the long-term technological needs of Africa. The promotion of science and technology as a way to meet human welfare needs must, however, take into account the additional need to protect Africa's environment for present and future generations.

The concept of "sustainable development" has been advanced specifically to ensure the integration of social, economic, and environmental factors in development strategies and associated knowledge systems.[12] Mapping out strategic options for Africa's economic renewal will therefore need to be undertaken in the context of sustainable development strategies and action plans.

There is widespread awareness of rapid scientific advancement and the availability of scientific and technical knowledge worldwide. This growth feeds on previous advances and inner self-propelling momentum. In fact, the spread of scientific knowledge in society is eroding traditional boundaries between scientists and the general public. The exponential growth in knowledge is also making it possible to find low-cost, high-technology solutions to persistent problems.

Life sciences are not the only areas where research could contribute to development. Two additional areas warrant attention. The continent's economic future crucially depends on the fate and state of its infrastructure, whose development will depend on the contributions of engineering, materials, and related sciences. It is notable that these fields are particularly underdeveloped in Africa and hence could benefit from specific missions that seek to use local material in activities such as road construction and maintenance. Other critical pieces involve expanding the energy base through alternative energy development programs. This sector is particularly important because of Africa's past investments, its available human resources, and its potential to stimulate complementary industries that provide parts and services to the expansion of the sector. Exploiting these opportunities requires supporting policies.

Advances in science and technology will therefore make it possible for humanity to solve problems that have previously existed in the realms of imagination. This is not a deterministic view of society but an observation of the growth of global knowledge and the feasibility of new technical combinations that are elicited by social consciousness. This view would lead to the conclusion that Africa has the potential to access more scientific and technical knowledge than the more advanced countries had in their early stages of industrialization.

Evidence shows the role that investment in research plays in Africa's agricultural productivity. For example, over the past three decades Africa's agricultural productivity grew at a much higher rate (1.8%) than previously calculated, and technical progress, not efficiency change, was the primary driver of this crucial growth.[13] This finding reconfirms the critical role of research and development (R&D) in agricultural productivity. The analysis also lends further support for the key role that pro-trade reforms play in determining agricultural growth.

Within North Africa, which has experienced the highest of the continent's average agricultural productivity growth (of 3.6% per year), Egypt stands out as a technology leader, as the gross majority of its agricultural growth has been attributable to technical investments and progress, not efficiency gains. A similar trend stressing the importance of technical progress and R&D has been seen in an additional 20 African countries that have experienced annual productivity growth rates over 2%.

This evidence shows that the adaptive nature of African agricultural R&D creates shorter gestation lags for the payout from R&D when compared to basic research. This makes the case for further investment even more important.[14] On the whole, African agricultural productivity increased incrementally from the 1970s to the 1980s to the 1990s. While growth in these decades can be attributed largely to major R&D investments made in the 1970s, declines in productivity growth in the 2000s are attributed to decreased R&D investments in the late 1980s and 1990s. With an average rate of return of 33% for

1970–2004, sustained investment in adaptive R&D is a crucial element of agricultural productivity and growth.

Evidence from other regions of the world tells the same story in regard to specific crops. For example, improvements in China's rice production illustrate the significant role that technical innovation plays in agricultural productivity. Nearly 40% of the growth in rice production in 13 of China's rice-growing provinces over the 1978–1984 period can be accounted for by technology adoption. Institutional reform could explain 35% of the growth. Nearly all the growth in the subsequent 1984–1990 period came from technology adoption. These findings suggest that the impact of institutional reform, though significant, has previously been overstated.[15] The introduction of new agricultural technologies went hand in hand with institutional reform.

Looking at production growth in crops such as maize, cotton, wheat, and oilseed, similar results were found. These crops had an average production growth of 4% a year in the period 1961–2004. The trends show that all crops except for oilseed grew rapidly in area, yield, and production between 1961 and 1980, a period during which new varieties, techniques such as irrigation, and inputs such as chemical fertilizers and pesticides were rapidly adopted.[16] South Korea underwent structural changes from 1961 to 2004. Large land area contractions contributed to negative annual production rates for wheat and soybeans. Yield increases in maize and rice contributed to positive growth.[17] In most cases, production growth was a result of increased yields, and production decreases were from contractions in crop area.

Platform Technologies

Information and Communication Technologies

While information and communication technologies in industrialized countries are well developed and historically established, ICTs in developing countries have traditionally been "based on indigenous forms of storytelling, song and theater, the print media and radio."[18] Despite Africa's current deficiency in more modern modes of communication and information sharing, the countries benefit greatly from the model of existing technologies and infrastructure.

In addition to the specific uses that will be explored in the rest of this section, ICTs have the extremely significant benefit of providing the means for developing countries to contribute to and benefit from the wealth of knowledge and research available in, for example, online databases and forums. The benefits of improved information and communication technologies range from enhancing the exchange of inter- and intra-continental collaborations to providing agricultural applications through the mapping of different layers of local landscapes.

One Kenyan company, Gro Intelligence, recognizes the important role of research and data in development, particularly in the agricultural sector.[19] An inadequate quantification of the sector's problems slows the development of their solutions. Agricultural data are particularly inadequate in Africa, despite the continent's unparalleled agricultural potential. To be of value, data need to be timely, accurate, and accessible. Currently, those interested in African agricultural data—whether related to production, prices, soil quality, or a range of other indicators—do not have a centralized database that offers high-quality data. And the consequences of this lacuna are significant: policymaking that is not evidence-based, for example, does not bring about desired changes, while still wasting funds. Investors that could bring improvements to infrastructure or credit access for smallholders are unwilling to enter opaque markets. Gro Intelligence is tackling this problem by building the world's first agricultural supercomputing platform. The Gro Intelligence Platform aggregates data, performs complex analytics, offers elucidatory visualizations, and layers in relevant sociopolitical and regulatory information.

Gro Intelligence adds value to the data through its analytics, offering correlation factors, predictive tools, and pricing analytics. Subscribers to the Gro Platform have access to precomputed data using proprietary algorithms and can also use proprietary Gro algorithms to run computations on demand.

Data sets related to African agriculture are messy, exist in different formats, and define terms differently. Gro Intelligence begins by aggregating all relevant data, which come from sources including government agencies, satellites, sensors, and nongovernmental organizations (NGOs). While aggregating this wide array of statistics (Gro currently tracks more than 40,000 indicators related to African agriculture), Gro makes adjustments that allow for the direct comparison of these dissimilar data sets, and places the information within its intuitive classification system. Gro also corrects errors, including missing values and abnormal outliers, while also prudently resolving any statistical conflicts.

These data are unique in their accessibility. Gro Intelligence offers customizable and easily comprehensible visual representations of various data sets. Visualizations include Gro Geo-Heat Maps, which use a multicolored matrix to represent magnitude, and Gro Choropleth Maps, which use shade intensity of a single color to represent proportion. Ultimately, Gro's platform seeks to deliver innovation with elegance in order to make the user experience as productive as possible. The Gro Intelligence Platform also gives users access to important contextual information relevant to the crops they are interested in, offering up-to-date news updates, insight into government policies, and original written analyses composed by Gro Intelligence.

The potential impact of this data is vast. Processors and logistics companies can get a real-time idea of where their services are needed. Policymakers will have the tools necessary to create relevant and powerful policies. Financial institutions, presently unable to offer smallholder farmers loans due to their inability to accurately measure risk, will be able to offer innovative financial products more appropriate for these farmers. Companies selling inputs, armed with better soil information, will be able to optimize the sale of fertilizer and seeds. And philanthropic groups and NGOs active in the sector can better measure and maximize their impact.

An improved understanding of African agricultural sectors will translate into an improved understanding of wider African economies. Data have the power to catalyze the transformation of agricultural markets and economies more broadly. In April 2014, Nigeria shocked the world when it announced that its newly rebased GDP was 89% larger than it was the previous year, meaning that it had overtaken South Africa as the continent's largest economy. Although many countries regularly reevaluate their GDP calculations, Nigeria's previous reevaluations held little weight given the fact that the country had not updated its national statistics since 1990. It is impossible to have an accurate understanding of an economy without building an accurate understanding of its agricultural sector.

Mobile Technology

Sub-Saharan Africa has 10 times as many mobile phones as landlines in operation, providing reception to approximately 80%[20] of the population. The mobile penetration rate (number of subscriptions per 100 inhabitants) in Africa is expected to reach 69% by the end of 2014.[21] Much of this growth in cell phone use—as much as 45% annually from 2002 through 2011—coincided with economic growth in the region. It is estimated that every 10% growth in mobile phones can raise up to 1.5% in GDP growth. There are five ways in which mobile phone access boosts microeconomic performance: reducing search costs and therefore improving overall market efficiency, improving productive efficiency of firms, creating new jobs in telecommunications-based industries, increasing social networking capacity, and allowing for mobile development projects to enter the market.[22]

Mobile phones cut out the opportunity costs, replacing several hours of travel with a two-minute phone call and also allowing firms and producers to get up-to-date information on demand. They also redistribute the economic gains and losses per transaction between consumers and producers. This reduction in the costs of information gathering creates an ambiguous net welfare gain for consumers, producers, and firms. Similarly, mobile phones make it easier for social networks to absorb economic shocks. Family and kinship relationships have always played an important role in African society, and mobile phones strengthen this already available "social infrastructure," allowing faster communication about natural disasters, epidemics, and social or political conflicts.

The use of more mobile phones creates a demand for additional employment. For instance, formal employment in the private transport and communications sector of Kenya rose by 110% between 2003 and 2011, as mobile phone use rose about 45% annually during that period. While there is a measurable growth in formal jobs, such as hotline operators who deliver information on agricultural techniques, there is also growth in the informal sector, including the sale of phone credit and "pay-as-you-go" phones, repair and replacement of mobile phone hardware, and operation of phone rental services in rural areas. New employment opportunities also come through the mobile development industry.

Africa's advantage over countries such as the United States in avoiding unnecessary infrastructure costs is especially exemplified in the prevalence of mobile technologies, which have replaced outdated landline connectivity. Mobile phones have a proven record in contributing to development, as illustrated by the associated rise in the rate of mobile phone use, averaging 65% annually over the last five years.[23] Because mobile phones are easy to use and can be shared, this mobility has revolutionized and facilitated processes such as banking and disease surveillance.

The potential and current uses of mobile technology in the agricultural sector are substantial and varied. For instance, local farmers often lacked the means to access information regarding weather and market prices, making their job more difficult and decreasing their productivity. With cellular phones comes cheap and convenient access to information such as the cost of agricultural inputs and the market prices for crops.

The desire for such information has led to the demand for useful and convenient mobile phone–based services and applications: "New services such as *AppLab*, run by the Grameen Foundation in partnership with Google and the provider MTN Uganda, are allowing farmers to get tailored, speedy answers to their questions. The initiative includes platforms such as *Farmer's Friend*, a searchable database of agricultural information, *Google SMS*, a question and answer texting service and *Google Trader*, a SMS-based 'marketplace' application that helps buyers and sellers find each other."[24] Simple services like text messaging have likewise led to an expansion of access and availability of knowledge. Applications such as these, coupled with the increased usage of cellular phones, have reduced the inefficiencies and unnecessary expenses of travel and transportation.

Internet Technology

From opening access to research institutes to facilitating business transactions, few technologies have the potential to revolutionize the African agricultural sector as much as the Internet. The demand for Internet service in Africa has been shown in the large increases in Internet usage (over 1,000% between 2000 and 2008) and through the fiber-optic cable installed in 2009 along the African east coast by Seacom. By 2011, four new undersea fiber optical cables were serving African western and eastern coasts. Just as with mobile phones, the Internet will have a transformative impact on the operations of businesses, governments, NGOs, farmers, and communities alike.

New statistical estimates from the International Telecommunication Union (ITU) show that close to 20% of the population in Africa will be using the Internet by the end of 2014, compared to 10% in 2010. However, only one in 10 households will have Internet access at home, compared to approximately 44% of households globally. The growth rate, however, is much larger in Africa than the rest of the world because of low penetration rates. More and more people will have access to Internet from their mobile phones. Mobile phone broadband penetration levels are close to 19% in Africa, which is up from 2% in 2010, showing high growth in a short time frame. Africa still lags when it comes to fixed-broadband subscriptions, with levels of 0.5%, compared to 44% in Asia and the Pacific.[25]

Until recently, Africa was served by an undersea fiber-optic cable only on the west coast and in South Africa. The rest of the continent relied on satellite communication. The first undersea fiber-optic cable, installed by Seacom, reached the east African coast in July 2009, as noted earlier. The US$600 million project has reduced business costs, created an e-commerce sector, and opened up the region to foreign direct investment.

New industries that create content and software are likely to emerge. This will in turn stimulate demand for access devices. A decade ago it cost more than US$5,000 to install one kilometer of standard fiber-optic cable. The price has dropped to less than US$300. However, for Africa to take advantage of the infrastructure, the cost of bandwidth must decline. Already, Internet service providers are offering more bandwidth for the same cost. By 2011, the four undersea cables operating in Africa had resulted in a quadrupling of data transfer speeds and a 90% price reduction.

Access to broadband is challenging Africa's youth to demonstrate their creativity and African leaders to provide a vision of the role of infrastructure in economic transformation. The emergence of Safaricom's M-PESA service—a revolutionary way to transmit money by mobile phones—is an indicator of great prospects for using new technologies for economic improvement.[26] In fact, these technologies are creating radically new industries, such as branchless banks, that are revolutionizing the service sector.[27]

Geographic Information Systems (GIS)

The diffusion of geographic information systems is creating new opportunities for development in general and Africa in particular, with regard to agriculture.[28] Several countries in Africa have invested in the development of space-based technologies through satellite programs in order to monitor land use and natural resources. The data provided by GIS technologies can play an important role for agriculture and food security as they give the opportunity to monitor land use, water catchment, crops and cropping systems, and for prospecting groundwater.[29]

An example from India shows the potential that GIS has for agriculture. Through the digitization of more than 20 million land records under the Bhoomi Project in India's State of Karnataka, the available information on land rights and land use innovation improved and became more available. But the implications of the Bhoomi Project did not just stop there. Because of the availability of such geospatial information, bankers became more inclined to provide crop loans and land disputes began declining, allowing farmers to invest in their land without fear. The success of this project has inspired the government of India to establish the National Land Records Modernisation Programme to do the same for the entire country.[30] Not only does this show the applicability and usefulness of information technologies in agriculture, but it also provides an option to be considered by African countries.

Unmanned aerial vehicles (UAVs) are another type of technology that the agricultural sector is starting to adopt. UAVs have the ability to collect local crop data and predict the use of inputs accordingly. They can also be used as vehicles to deliver herbicides and pesticides to the crops. Proper monitoring of what the crop needs, together with the right inputs, has the potential to sustainably intensify the agricultural sector and to reduce the use of inefficient inputs and crop damage. UAVs can also be used for risk management by mapping crop levels, water levels, and overall weather patterns.[31] The Rwandan government has embarked on the introduction of modern technology to improve production and improve food security, in line with its Vision 2020, which highlights agriculture as one of the six pillars of economic growth. The main objective is to implement modern practices and technologies to improve production capabilities by developing a reliable and cost-effective method to monitor agriculture and introduce precision farming in Rwanda. Currently the Ministry of Agriculture and Animal Resources is focusing on how UAVs can be used to modernize the sector. Specifically, it is examining how UAVs can be used to collect aerial data, which can in turn be used in a decision support system for modeling crop yield.

One notable problem the Ministry faces is in continuous monitoring to assess immediate and long-term impact of these practices. In 2008 the Ministry collected aerial imagery of the entire country; however, high cost and painstaking planning means that aerial photography is not the most reliable method for continuous monitoring. UAVs, on the other hand, can provide high-quality data at an affordable price compared to aerial imagery or very high-resolution satellite data, and they can provide the real-time and frequent data collection that is essential to crop monitoring. One important use of this data is for land monitoring, which includes land management practices that can help reduce degradation and soil erosion. Second, it can be used to enhance precision agriculture, which can help farmers vary the rate of input use in order to reduce costs and improve efficiency. Third, one of the most important applications is using decision support systems for agro-technology transfer to produce yield maps, which are created by integrating data from

UAVs as well as on-the-ground data and can be made available to all farmers through smartphone applications.

Unlike biotechnology and nanotechnology, ICT and GIS do not have as much risk of being overregulated or reviled for being a great unknown. However, regulation of ICT and GIS will be necessary—and keeping clients and users secure will be a challenge as more and more of Africa becomes connected to the international network. Balancing privacy with the benefits of sharing knowledge will probably be one of the largest challenges for these sectors, especially between countries and companies.

When introducing new technologies in the field, policymakers should consider ones that are low cost and easily accessible to the farmers who will use them. They should ideally capitalize on techniques that farmers already practice and involve support for scaling up and out, rather than pushing for expensive and unfamiliar practices.

New technologies should also allow farmers to be flexible, according to their own capacities, situations, and needs. They should require small initial investments and should let farmers experiment with the techniques to decide their relative success. If farmers can achieve success with a small investment early on, they are likely to devote more resources to the technique later as they become committed to the practice. An example of this type of technological investment is the planting pits that increased crop production and fostered more productive soil for future years of planting.

Nanotechnology

Of the platform technologies discussed in this chapter, nanotechnologies are the least explored and most uncertain. Nanotechnology involves the manipulation of materials and devices on a scale measured by the billionths of a meter. The results of research in nanotechnology have produced substances of both unique properties and the ability to be targeted and controlled at a level unseen previously. Thus far, applications to agriculture have largely been theoretical, but practical projects have already been explored by both the private and public sectors in developed, emerging, and developing countries.[32]

For example, research has been done on chemicals that could target one diseased plant in an entire crop. Nanotechnology has the potential to revolutionize agriculture with new tools such as the molecular treatment of diseases, rapid disease detection, and enhancement of the ability of plants to absorb nutrients. Smart sensors and new delivery systems will help to combat viruses and other crop pathogens. Increased pesticide and herbicide effectiveness, as well as the creation of filters for pollution, create more environmentally friendly agriculture processes. While countries like the United States and China have been at the forefront of nanoscience research expenditure and publications, emerging countries have engaged in research on many of the applications, from water purification to disease diagnosis.

Water purification through nanomembranes, nanosensors, and magnetic nanoparticles have great, though currently cost-prohibitive, potential in development, particularly in countries like Rwanda, where contaminated water is the leading cause of death. However, the low energy cost and high specificity of filtration has led to a push for research in water filtration and purification systems like the Seldon WaterStick and WaterBox. Developed by US-based Seldon Laboratories, these products require low energy usage to filter various pathogens and chemical contaminants and are already in use by aid workers in Rwanda and Uganda.[33]

One of the most promising applications of nanotechnology is low-cost, energy-efficient water purification. Nearly 300 million people in Africa lack access to clean water. Water purification technologies using reverse osmosis are not available in much of Africa, partly because of high energy costs. Through the use of a "smart plastic" membrane, the US-based Dais Analytic Corporation has developed a water purification system that could significantly increase access to clean water and help to realize the recent proclamation by the United Nations that water and sanitation are fundamental human rights.

The capital costs of the NanoClear technology are about half the cost of using a reverse osmosis water purification system. The new system uses about 30% less energy and does not involve toxic elements. The system is modular and can be readily scaled up on demand. A first-generation pilot plant opened in Tampa, Florida, in 2010, to be followed later in the year by the deployment of the first fully operational NanoClear water treatment facility in northern China. The example of Nano-Clear illustrates how nanotechnology can help provide clean water, reduces energy usage, and charts an affordable course toward achieving sustainable development goals.[34]

The potential for technologies as convenient as these would revolutionize the lifestyles of farmers and agricultural workers in Africa. Both humans and livestock would benefit from disease-free, contaminant-free water for consumption and agricultural use.

The cost-prohibitive and time-intensive process of diagnosing disease promises to be improved by nanotechnological disease diagnostics. While many researchers have focused on human disease diagnosis, developed and developing countries alike have placed an emphasis on livestock and plant pathogen identification in the interest of promoting the food and agricultural industry. Nanoscience has offered the potential of convenient and inexpensive diagnosis of diseases that would otherwise require time and travel. In addition, the convenient nanochips would be able to quickly and specifically identify pathogens with minimal false diagnoses. An example of this efficiency is found in the EU-funded Optolab Card project, whose kits allow a reduction in diagnosis time from 6–48 hours to just 15 minutes.

Another nanotechnology that could be used in agriculture is biopolymers, which can encapsulate fertilizer, contributing to slow release in the soil. These biopolymers are made of organic material such as cellulose. A common problem with some types of fertilizer application is the loss of nutrients because the plant is not able to absorb all the fertilizer given at once. The excess fertilizer that is not absorbed by the plant can damage the soil as well as release greenhouse gases into the atmosphere. This technology could therefore reduce the sector's dependence on fossil fuels and increase the effectiveness of the fertilizers. Coated fertilizers with modified membranes can also be tailored to address local crops and conditions. Biopolymers can also be used to reduce the stream of plastic waste from agriculture and food processing by enhancing the use of bioplastics that achieve the same function but halt the spread of plastics into the environment. Finally, biopolymers can provide safe and efficient methods of combating water scarcity and fluctuation through the development of biodegradable water management soil additives.

In 2011, COMESA recognized this opportunity and launched an initiative to support the development of a biopolymer and biomaterials cluster platform. The underlying idea was that the creative use of biopolymers and biomaterials can help empower COMESA countries to utilize their rich agricultural resources to foster green growth strategies and can offer solutions for challenges such as sustainable agriculture, health, water, energy, and environmental management. Then, in January 2013, Jomo Kenyatta University of Agriculture and Technology (JKUAT) hosted the first African workshop on biopolymers. The goal of the workshop was to establish a frontier biopolymer research program for Africa, with a focus on current advancements in biopolymer research with applications for industry, public policy, and international science and technology cooperation. The second African biopolymer workshop, held in May 2013 at the University of Mauritius, stressed the need to strengthen national innovation systems.

JKUAT is now working with the Kenyan Ministry of Industrialization to establish the African Biopolymer Center of Excellence, which will provide the infrastructure needed to carry out a program of research as outlined in the first workshop. That program will include bioresource utilization, biopolymer utilization in health, waste management, and environmental protection, and biopolymer utilization in agriculture as described above. Transferring the production know-how to farmers and the industry is another key component of the Center.

Technology Prospecting

Much of the debate on the place of Africa in the global knowledge economy has tended to focus on identifying barriers to accessing new technologies. The basic premise has been that industrialized countries continue to limit the ability of developing countries to acquire new technologies by introducing restrictive intellectual property rights. These views were formulated at a time when technology transfer channels were tightly controlled by technology suppliers, and developing countries had limited opportunities to identify the full range of options available to them. In addition, they had limited capacity to monitor trends in emerging technologies. But more critically, the focus on new technologies, as opposed to useful knowledge, hindered the ability of developing countries to create institutions that focus on harnessing existing knowledge and putting it to economic use.

In fact, the Green Revolution and the creation of a network of research institutes under the Consultative Group on International Agricultural Research (CGIAR) represented an important example of technology prospecting. Most of the traits used in the early breeding programs for rice and wheat were available but needed to be adapted to local conditions. This led to the creation of pioneering institutions such as the International Maize and Wheat Improvement Center (CIMMYT) in Mexico and the International Rice Research Institute (IRRI) in the Philippines.[35]

Other countries have used different approaches to monitor, identify, and harness existing technologies, with a focus on putting them to commercial use. One such example is the Fundación Chile, established in 1974 by the country's Minister for Economic Cooperation, engineer Raúl Sáez. The Fundación Chile was set up as joint effort between the government and the International Telephone and Telegraph Corporation to promote research and technology acquisition. The focus of the institution was to identify existing technologies and match them to emerging business opportunities. It addressed a larger goal of helping to diversity the Chilean economy and created new enterprises based on imported technologies.[36]

Unlike their predecessors, who had to manage technological scarcity, Africa's leaders today face the challenge of managing an abundance of scientific and technological knowledge. The rise of the open access movement and the growing connectivity provided by broadband Internet now allow Africa to dramatically lower the cost of technology searches. But such opportunities require different technology acquisition strategies. First, they require the capacity to access the available knowledge before it becomes obsolete. Second, such assessments have to take into account the growing convergence of science and technology.[37] There is also an increasing convergence between different disciplines.

Moreover, technology assessments must now take into account social impacts, a process that demands greater use of the diverse disciplines.[38] Given the high rate of uncertainty associated with the broader impact of technology on environment, it has become necessary to incorporate democratic practices such as public participation in technology assessments.[39] Such practices allow the public to make necessary input into the design of projects. In addition, they help to ensure that the risks and benefits of new technologies are widely shared.

Reliance on imported technology is only part of the strategy. African countries are just starting to explore ways to increase support for domestic research. This theme should be at the center of Africa's international cooperation efforts.[40] These measures are an essential aspect of building up local capacity to utilize imported technology. This insight is important because the capacity to harness imported technology depends very much on the existence of prior competence in certain fields. Such competence may lie in national research institutes, universities, or enterprises. The pace of technology absorption is likely to remain low in countries that are not making deliberate efforts to build up local research capacity, especially in the engineering sciences. One way to address this challenge is to start establishing regional research funds that focus on specific technology missions.

Conclusion

The opportunities presented by technological abundance and diversity, as well as greater international connectivity, will require Africa to think differently about technology acquisition. It is evident that harnessing existing technologies requires a more detailed understanding of the convergence between science and technology as well the various disciplines. In addition, it demands closer cooperation between the government, academia, the private sector, and civil society in an interactive process. Such cooperation will need to take into account the opportunities provided by the emergence of Regional Economic Communities as building blocks for Africa's economic integration. All the RECs seek to promote various aspects of science, technology, and innovation in general and agriculture in particular. In effect, it requires that policymakers as well as practitioners think of economies as innovation systems that evolve over time and adapt to change.

3

LEAPFROGGING IN GENETIC TECHNOLOGIES

The role of agricultural biotechnology—more specifically, trans-genic crops—is one of the most controversial themes in African agriculture. This controversy exists despite the limited number of countries growing such crops. Much of the debate over the crops has been framed in political terms forcing individuals, organizations, and nations to take generic positions that do not reflect the case-by-case implications of transgenic crops.

This chapter offers an alternative approach to assessing the relevance of transgenic crops in African agriculture by focusing first on the frontiers of technologies. It stresses the importance of focusing on emerging applications that have the potential to maximize the benefits of the technology while reducing its risks. Next it examines the current trends. Third, the chapter looks at the benefits and risks of biotechnology. It concludes by explaining the importance of adjusting regulations to reflect advances in technology. Taking this approach requires policymakers to abandon the traditional juxtaposition of opponents and advocates of the technology and to use evidence for making decisions.

Frontiers of Agricultural Biotechnology

There can be little doubt that improved crop varieties are necessary to feed a growing population, mitigate climate change, and minimize agriculture's large environmental footprint.

Crop varieties are improved through conventional breeding methods such as cross-breeding or hybridization, or through more sophisticated techniques such as genetic modification (GM). To be clear, these are all good options. Choosing the optimal method depends on the context. All of these methods should be made available to farmers so that they can choose the technique that best suits their needs. Regardless of the method, the goal of plant breeding remains the same: to select traits to create useful new varieties. Genetic modification allows plant scientists to introduce additional variations that cannot be done through conventional breeding. Because of the controversy surrounding genetically modified, or transgenic, crops, the benefits of this technique are often overlooked. The rest of this chapter will show how African countries can adopt second-generation GM technology and reap the benefits that have accrued to other adopters.

New techniques are currently pushing the boundaries of genetic plant breeding. As synthetic biology advances, genetic modification is also likely to continue to improve as new genes and traits are found. Second-generation technology now allows for stacked traits, transfer of one or more specific genes into an exact location in the "recipient plant genome," and new traits that "increase photosynthetic efficiency, nitrogen use efficiency, aluminum tolerance, salinity tolerance and phosphate use efficiency in crop plants." Improvements in synthetic biology will drastically simplify current plant-breeding methods. For example, using synthetic biology, scientists can insert multiple genes into a plant at one specific location that will "confer multiple traits (herbicide, insect, disease, stress resistance, and nutritional enhancement)."[1]

GM is not the only technical method for enhancing crops varieties. Genetic engineering is arguably the newest and most encouraging frontier of agricultural biotechnology. Crop breeding through biotechnology is much broader than genetic modification. Gene editing techniques are especially promising because they eschew the most controversial aspects of genetic modification by "tweaking" or editing existing plant DNA to change the amount of natural ingredient already present in a particular crop, instead of inserting new ones from foreign organisms. Gene editing allows synthetic biologists to remove sources of unwanted DNA. They can also edit or adjust existing DNA to enhance certain endogenous features. As a natural process, genetic engineering offers an overarching strategy to make agriculture sustainable.

These techniques will be especially useful in the early stages in breeding improved fruit crop varieties, but their scope of applications will expand with time. Conventional fruit breeding is a slow, tedious process. Marker-assisted breeding helped to some degree, but genetic engineering techniques offer a higher degree of control in manipulating the plant genome than was possible before.[2]

As the genetic codes of more crops are sequenced, and as new DNA-editing techniques become more common, it will become much easier to enhance certain desirable traits such as sweetness, pest resistance, and disease resistance in crops. It is likewise suggested that these crops could be considered non-transgenically altered since they would not contain any foreign DNA.[3] This, however, will depend on regulators' interpretation of the new gene-editing techniques. Argentina, the European Union, the United States, Australia, and New Zealand already are considering new regulations for genetically edited crops.[4]

Trends in Biotechnology

Sub-Saharan Africa is well poised to benefit from those advantages that accrue to late adopters of new technologies. These countries can implement innovation systems that are unconstrained by the "lock-in" that developed countries face in numerous industries.[5] Sub-Saharan Africa also has powerful competitive advantages based on an abundance of natural resources, such as sunlight and underutilized arable land, that can help countries reach their full agricultural production potential. Indeed, agriculture is a promising sector in which countries can leapfrog into sustainable technologies. Countries that adopted less expensive second-generation biotechnology, for example, have experienced advantages that eluded early adopters. As the rest of this chapter will show, transgenic crops in particular offer sub-Saharan Africa the chance to adopt and benefit from second-generation biotechnology and to leapfrog into more sustainable methods of agricultural production. Countries cannot benefit from leapfrogging, however, without their own research capabilities in all aspects of the value chain. They must be able to take advantage of technological abundance, especially in agricultural biotechnology, by building on existing knowledge and adapting it to address local issues.

Biotechnology—technology applied to biological systems— has the promise of leading to increased food security and sustainable forestry practices, as well as improving health in developing countries by enhancing food nutrition. The uptake of transgenic crops is the fastest adoption rate of any crop technology, increasing from 1.7 million hectares in 1996 to 181.5 million hectares in 2014, and a 100-fold increase over the period.[6] By adopting agricultural biotechnology now, Africa has an opportunity to benefit from the latecomer advantage.

Recent increases among early adopting countries have come mainly from the use of "stacked traits" (instead of single traits in one variety or hybrid). In 2013, for example, 90% of the 39.4 million hectares of maize grown in the United States was transgenic, and 71% of this involved hybrids with double or triple stacked traits. Nearly 90% of the cotton growth in the United States, Australia, and South Africa is transgenic and, of that, 75% has double-stacked traits.

In 2014, there were 18 million farmers growing transgenic crops in 28 countries around the world, of whom over 90% (16.9 million) were smallholder farmers from developing countries. Notably, for the third year in a row, developing countries collectively planted more hectares of transgenic crops (96.2 ha) than industrial countries (85.3 ha). Most of the benefits to such farmers have come from cotton. In 2014, India achieved record production levels on 11.6 million hectares of *Bt* cotton (an estimated 40 million bales, up from 39 million in 2013). Remarkably, in 2014 India became the number one cotton-producing country in the world, overtaking China. Over the 2002–2013 period, *Bt* cotton added US$16.7 billion worth of value to Indian farmers, cut insecticide use by half, helped to double yield, and turned the country from a cotton importer into a major exporter.[7]

Africa is steadily joining the biotechnology revolution. Collectively, Burkina Faso, Sudan, and South Africa planted more than 3.3 million hectares in 2014. South Africa's transgenic crop production stood at 2.7 million hectares in 2014. In 2012–2013, Burkina Faso produced 630,000 tons of *Bt* cotton, up from 400,000 tons in 2011–2012. In 2013, South Sudan increased its production of *Bt* cotton by 300%.[8]

African countries, by virtue of being latecomers, have had the advantage of using second-generation transgenic seed. Monsanto's Genuity™ Bollgard II® (second generation) cotton contains two genes that work against leaf-eating species such as armyworms, budworms, bollworms, and loopers. They also protect against cotton leaf perforators and saltmarsh caterpillars. Akin to the case of mobile phones, African farmers can take advantage of technological leapfrogging to reap high returns from transgenic crops while reducing the use of chemicals. In 2013, 10 countries conducted confined field trials of various transgenic crops, including Burkina Faso (*Bt* cowpea); Cameroon (*Bt* cotton); Egypt (drought and salt-tolerant wheat); Ghana (*Bt* cotton, *Bt* cowpea, NUWEST rice, and high-protein sweet potato); Kenya (drought-tolerant and IR maize, IR cotton, *Bt* cassava, high-protein sweet potato, enhanced sorghum, IR pigeon pea, and Gypsophila flower); Malawi (*Bt* cotton); Nigeria (*Bt* cowpea); South Africa (drought-tolerant maize, stacked IR/HT cotton, and modified soybean); Sudan (IR cotton);

and Uganda (drought-tolerant and IR maize, *Bt* banana, cassava, and NUWEST rice). Of these, Cameroon, Ghana, Kenya, Malawi, Nigeria, and Uganda have ongoing field trials. The field trials indicate that many of these crops are second-generation. The next step for those countries is commercialization. The door is now open for the revolutionary adoption of biotechnology that will extend to other crops as technological familiarity and economic benefits spread.

There is also a rise in the adoption of transgenic crops in Europe. In 2011, five European countries—Czech Republic, Slovakia, Portugal, Romania, and Spain—planted commercial *Bt* maize, led by Spain with 136,962 hectares. Trends in Europe suggest that future decisions on transgenic crops will be driven by local needs as more traits become available. For example, crops that tolerate various stresses such as drought are likely to attract interest among farmers in Africa.

These trends also demonstrate new efforts by leading global research firms to address the concerns of resource-poor farmers, a sub-theme in the larger concern over the contributions of low-income consumers.[9] Other traits that improve the efficiency of nitrogen uptake by crops will also be of great interest to resource-poor farmers. Other areas that will attract interest in developing new transgenic crops will include the recruitment of more tree crops into agriculture and the need to turn some of the current grains into perennials.[10]

Trends in regulatory approvals are a good indicator of the future of transgenic crops. By 2014, some 28 countries had planted commercial transgenic crops and another 31 had approved transgenic crop imports for food and feed use and for release into the environment. A total of 1,045 approvals have been granted for 196 events (unique DNA recombinations in one plant cell used to produce entire transgenic plants) for 25 crops. Transgenic crops are accepted for import in 60 countries (including Japan, the United States,

Canada, South Korea, Mexico, Australia, the Philippines, the European Union, New Zealand, and China). The majority of the events approved are in maize (65), followed by cotton (39), canola (15), potato (14), and soybean (14).[11]

Benefits of Biotechnology

Technology helped to generate significant increases in agricultural productivity during the Green Revolution. The combination of new high-yielding crop varieties, agro-chemicals, and better irrigation techniques raised food production among farmers in developed and developing countries exponentially. As a result, poverty and hunger declined throughout much of Asia as food levels rose, prices fell, and trade and consumption increased.

The favorable conditions that led to the success of the Green Revolution, however, have changed. Staple crops will be most affected by the "exhaustion of some past sources of growth [making] future yield expansion as great a challenge as in the past."[12] Overuse of fertilizers and chemical pesticides has led to pest and weed resistance and environmental degradation. Moreover, the overall availability of arable land is declining, water resources are scarce, and climate change is causing significant changes in weather patterns, all of which makes it necessary to find alternatives to current production methods. Transgenic crops are one alternative to addressing these challenges, as they are specifically designed to increase production while decreasing the use of chemicals, allowing late adopters to leapfrog the problems associated with Green Revolution technologies. There are now robust studies published that confirm the benefits of transgenic crops, including production and economic benefits, nutritional benefits, and environmental benefits.[13]

A key point, however, is that transgenic crops were not developed to increase yield directly but instead to remove barriers to production such as diseases, weeds, or pests. Increased production is necessary to meet an ever-increasing demand for food. The transgenic soybean enabled double-cropping in Argentina, which specifically met the huge increase in soy demand—driven primarily by an increased desire for meat in Asia—with only a limited effect on prices.[14]

Transgenic Crops and Food Security

Despite the high adoption rate of transgenic crops, their benefits are still highly debated. Meanwhile, around 870 million people are undernourished. To address these issues, current food production levels must be doubled. There are many claims that biotechnology cannot contribute to solving food insecurity or benefit smallholder farmers. These claims are driven by a wide range of concerns that tend to assert what has not been denied and to deny what has not been asserted. Transgenic crops have the potential to increase agricultural productivity on existing arable land; address issues of loss; increase access to food through income gains; raise nutrition levels; and promote sustainable agriculture. But realizing the potential needs to be viewed in a wider food security context, as well as from the lens of the latecomer advantage.

In fact, transgenic crops can benefit smallholder farmers in several major ways. Most importantly, they help farmers avoid both production and income loss due to pests, disease, and environmental factors such as drought or flooding. This results in greater productivity. Insect-resistant traits are found to have the greatest impact in warm, tropical places where pests are more prevalent and where insecticides and inputs are not widely used—namely, in emerging countries. For example, because of pest attacks, cotton was, until the early 1990s, the target of 25% of worldwide insecticide use.[15] Recombinant DNA engineering of a bacterial gene that codes for a toxin lethal to bollworms resulted in pest-resistant cotton, increasing profit and yield while reducing pesticide and management costs.[16]

Countries such as China took an early lead in adopting the technology and have continued to benefit from the reduced use of pesticides.

Food security is about expanding ecologically sustainable agricultural practices as well as increasing access to nutritious food. Biotechnology can play a role in increasing agricultural productivity, income levels, nutrition, and stability and resilience of the food system to various shocks, thereby helping to increase food security at the global level but especially in emerging countries.

Production and Economic Benefits

Recent studies of transgenic cotton in India and China confirmed earlier results: transgenic cotton production per hectare is demonstrably higher than that of nontransgenic cotton, especially in India. India went from having one of the lowest rates of cotton production to a 24% increase in yield and a 50% increase in profit among smallholder farmers after 90% of its farmers adopted transgenic cotton.[17] Other benefits include decreased pesticide use especially in China, and health benefits in both countries. Cotton was the most adopted transgenic crop globally and saw the highest production increase. The global price effects of planting *Bt* cotton are estimated at 10%.[18] This in turn raised farmers' income and consumption levels, allowing them to buy other foodstuff, thus increasing food security.

Although there are concerns that cotton production levels will decline over time as a result of bollworm resistance, the evidence so far does not support these concerns, as "aggregate cotton yields continue to rise in China suggesting that *Bt* cotton also continues to do well."[19]

A global impact study confirms the income gains among Indian and Chinese farmers who adopted transgenic IR cotton, transgenic *Bt* soybeans in South America, and a variety of transgenic crops in the United States. South Africa, the

Philippines, Mexico, and Colombia are also seeing the income benefits of adopting transgenic crops. These gains stem from greater productivity and efficiency. Some of the largest income gains derive from the maize sector. In fact, "$6.7 billion additional income generated by transgenic insect resistant maize in 2012 has been equivalent to adding 6.6% to the value of the crop in the transgenic crop growing countries, or adding the equivalent of 3% to the $226 billion value of the global maize crop in 2012. Cumulatively since 1996, transgenic IR technology has added $32.3 billion to the income of global maize farmers."[20] Furthermore, planting transgenic crops can reduce input expenses associated with pesticide use, such as machinery costs, fuel costs, and water use. Although seed prices for transgenic cotton are often higher than for conventional seeds in India, these costs are offset.

In Africa, where smallholder farmers use significantly fewer inputs than in developed countries, IR crops could have a large impact on production. By adapting the technology to local conditions, developing countries could also address the issue of yield drag, which occurs when companies typically modify generic seeds that are unspecific to a particular region. African countries could increase the production potential of transgenic crops by applying the technology to high-quality, local crop varieties. This can help reduce loss from pests, weeds, and diseases. The potential of this technology lies in how it is adapted to meet specific, local needs in developing countries, which can range from combating diseases to improving indigenous crops.

For example, tissue culture of bananas has had a great impact on the economy of East African countries since the mid-1990s. Because of its susceptibility to disease, banana has always been a double-edged sword for African economies like that of Uganda, which consumes a per capita average of one kilogram per day. Nearly 15 million people rely on bananas for their income or consumption, making it one of the most important crops in Uganda. For example, when the Black Sigatoka fungus arrived in East Africa in the 1970s, banana productivity decreased as much as 40%. Tissue culture experimentation allowed for quick generation of healthy plants and was met with great success. Since 1995, Kenyan banana production has more than doubled, from 400,000 to over one million tons in 2004, with average yield increasing from 10 tons per hectare to 30–50 tons.

Researchers in Uganda are currently using biotechnology to reverse the trend of *Xanthomonas* wilt, a bacterial disease that causes discoloration and early ripening of bananas and costs the Great Lakes region approximately US$500 million annually. There is currently no treatment for the disease, and given the crop's status as a staple in this region, solving this problem would directly increase food security and income.[21] The most efficient method of containing the disease is by growing transgenic bananas instead of relying on more labor-intensive methods of removing and destroying affected bananas. By transferring two genes from green peppers, scientists were able to grow highly resistant bananas. Results from field trials in Uganda and Kenya are extremely promising, but the regulatory regimes do not yet allow for commercialization.

In Nigeria the insect *Maruca vitrata* destroys nearly US$300 million worth of blackeyed peas—a major staple crop—and forces farmers to import pesticides worth US$500 million annually. To solve the problem, scientists at the Institute for Agricultural Research at Nigeria's Ahmadu Bello University have developed a pest-resistant, transgenic blackeyed pea variety using insecticide genes from the *Bacillus thuringiensis* bacterium. The crop is also undergoing field trials in Burkina Faso and Ghana, and is slated for release to farmers in 2017.[22]

In Southeast Asian countries such as Bangladesh, India, and the Philippines, *Bt* brinjal is the region's first transgenic food crop and offers economic, nutritional, and environmental benefits. Researchers and scientists at the Bangladesh Agricultural Research Institute developed *Bt* brinjal to resist the "fruit and shoot borer," with support from USAID and Cornell

University. The result was significantly fewer pesticide sprays during the growing period and fewer dips in pesticide just before harvest. The transgenic eggplant has obvious farmer health and environmental benefits from reduced pesticide use. The crop was commercialized in Bangladesh, but its future remains controversial as the government and opponents of transgenic crops seek to push or stall further crop sales. Furthermore, the Filipino government prohibited field trials of *Bt* brinjal, citing health and environmental concerns. As a result, commercialization of the crop remains stalled in India and the Philippines.[23]

It is also important to note what is not in the pipeline, namely smaller crops that are staples in certain regions of the world but are unlikely to be developed in the foreseeable future because of prohibitive regulatory costs and risks. Regardless, promising transgenic vegetable crops such as insect-resistant bananas, blackeyed pea, eggplant, papaya, sweet corn, summer squash, plums, citrus fruits, and wheat must clear significant resistance and regulatory hurdles before their societal benefits can be realized.

Nutritional Benefits

Interest in transgenic crops also includes their potential contribution to nutritional enhancement in staple crops, specifically targeting low-income families.

There are several bio-fortified crops that are currently available or being tested in developing countries. Two of the most promising include "Golden Rice," containing more beta carotene or Vitamin A, under evaluation in the Philippines and Bangladesh; and the "Golden Banana," developed by Ugandan researchers and bio-fortified with Vitamin A and iron. Here, scientists applied the pro-Vitamin A genes used in Golden Rice to a popular local crop to help solve a regional health issue. Addressing vitamin deficiencies would lead to lower healthcare costs and higher economic performance.

In the United Kingdom, researchers at the John Innes Centre created a bio-fortified "purple tomato" by expressing genes from the snapdragon in the transgenic tomato. The dark color derives from the same antioxidant that is found in blueberries and cranberries—anthocyanin—and offers similar health benefits at a lower cost to consumers. By increasing the antioxidant levels in a common food such as the tomato, researchers hope to stimulate greater consumption of antioxidants. The purple tomato contains the "highest levels of anthocyanins yet reported in tomato fruit," and an early study of cancer-prone rats suggests that the tomato's high levels of anthocyanins increased the life span of these rats when eaten regularly. The purple tomato also has a longer shelf life than a nontransgenic tomato.[24]

Nutritional enhancements through genetic modification are still in their infancy. Examples such as Golden Rice, Golden Bananas, and purple tomatoes are important because they represent proof of concept. Once they are confirmed, they will open a wide range of opportunities for related modifications in other crops, as well as the use of new techniques to improve human nutrition.

Environmental Benefits

Climate change will adversely affect agricultural productivity primarily in developing countries. Many regions are expected to suffer production loss as drought, floods, storms, changes in sea level, and overall warmer temperatures become more common. Developing countries will bear the brunt of these changes. In the past, farmers and regions could recover from these relatively rare events during the next growing season. Now it is imperative to determine ways of increasing the resilience and stability of food systems so that productivity is less affected by drought, flood, or both in the same season. Challenges include increasing productivity on existing land to conserve biodiversity and protect vulnerable land, as well as reducing agriculture's large environmental footprint.

Transgenic crops are actually one of the better land-saving technologies available, as they are designed to increase production on existing plots, avoiding the slash and burn agriculture often practiced in developing countries. Indeed, "if the 377 million tons of additional food, feed and fiber produced by biotech crops during the period 1996 to 2012 had been grown conventionally, it is estimated that an additional 123 million hectares . . . of conventional crops would have been required to produce the same tonnage."[25]

Transgenic crops have also succeeded in reducing pesticide use by 8.5% in 2011 alone and reducing fossil fuels and CO_2 emissions through less plowing and less chemical spraying (in 2012 the planting of transgenic crops helped reduce CO_2 emissions by approximately 26.7 billion—the equivalent of removing 11.8 million cars from the road).

In 2012, drought wreaked havoc on maize production in the United States, highlighting what farmers in Africa already know: drought is one of the biggest constraints to agricultural production worldwide. The development of drought-tolerant crops is one of, if not the, most important transgenic traits that will be commercialized over the next decade. The gene in question was isolated from a common soil bacterium known as *Bacillus subtilis*. It helps the plant cope better with stress caused by water shortages, allowing the plant to focus on filling the grains. In 2013, some 2,000 American farmers started to grow drought-tolerant maize. Indonesia has approved field trials of drought-tolerant sugarcane. Field trials of drought-tolerant maize, wheat, rice, and sugarcane are in field trials in Argentina, Brazil, India, Egypt, South Africa, Kenya, and Uganda. It is hoped that the first drought-tolerant maize will be commercially available in sub-Saharan Africa by 2017.

In March 2008, a public-private partnership called Water Efficient Maize for Africa (WEMA) was formed between Monsanto, which developed the drought-resistant technology;

the African Agricultural Technology Foundation, which directs the partnership; the International Maize and Wheat Improvement Center (CIMMYT); and five national agricultural research systems in East and Southern Africa (Kenya, Mozambique, South Africa, Tanzania, and Uganda). WEMA is working to make the drought-resistant technology available to smallholder farmers through local and regional seed companies. The crop is being developed using conventional breeding, marker-assisted selection, and genetic modification to find the optimal crop for local conditions. Confined field trials thus far show 20%–30% higher production than conventional hybrids. Sites were selected specifically for their dry conditions. The five national research systems are coordinating the field trials. WEMA hopes to offer at least five "farmer-preferred" IR maize hybrids with and without the drought-tolerant gene by 2017, pending field trials and regulatory approval. It is undergoing field trials in Kenya, South Africa, and Uganda, but the regulatory regimes in Mozambique and Tanzania so far prohibit field trials.

Furthermore, 25% of the global rice supply comes from flood-prone regions. One solution has been to isolate the gene present in a variety of Indian rice that allows plants to survive after up to three weeks underwater. In collaboration with the International Rice Research Institute (IRRI), researchers at the University of California at Davis used marker-assisted selection to breed this gene into locally important varieties. The result is a variety of rice that can tolerate flooding but which also retains the capability to produce at a high rate. IRRI partnered with PhilRice, a nonprofit organization in the Philippines, to distribute the rice free of charge to seed growers and certain farmers who can disseminate further to other farmers. In 2011, over 1 million farmers in the Philippines, Bangladesh, and India planted the rice. Other varieties are also being studied, including drought tolerance, heat and cold tolerance, and salt tolerance. In Africa, IRRI is partnering with the Africa Rice Center (AfriRice) to develop rice that can tolerate poor soils.

Two other crops in the pipeline are being developed to resist cold temperatures (eucalyptus) and drought (sugarcane). Several other techniques discussed earlier, including genetic engineering, tissue culture, diagnostics, genomics, and marker-assisted selection, can be used collectively to isolate new traits such as drought or flood tolerance. These examples prove that agricultural biotechnology has the potential to mitigate the effects of climate change and increase the resilience of food systems worldwide.

Risk Management and Public Policy

The safety of transgenic foods for human consumption and their impact on the environment has been a hotly debated issue. It gained international prominence following the publication of a paper that claimed that transgenic maize containing *Bt* genes caused cancer in rats.[26] The paper was used as a basis for regulatory action against transgenic foods in a number of countries. Upon closer scrutiny, however, several regulatory bodies, including the European Food Safety Agency, condemned the study as being methodologically defective.[27] The paper was later retracted by the journal that published it.

It is important to apply a case-by-case approach and to focus on foods that are on the market. Detailed reviews of the evidence so far available have come to the conclusion that the transgenic foods currently on the market carry the same risk profile as their conventional counterparts.[28] A comprehensive review of safety studies published over the last decade has examined the available evidence on the "safety of the inserted transgenic DNA and the transcribed RNA, safety of the protein(s) encoded by the transgene(s) and safety of the intended and unintended change of crop composition."[29] While acknowledging the need for further research, the review confirmed the general understanding that transgenic foods on the market today do not carry unique risks.

Building Research Capacity

Developing countries also face a separate set of risks from those of industrialized countries. For example, new medicines could have different kinds and levels of effectiveness when exposed simultaneously to other diseases and treatments. Similarly, "new technologies may require training or monitoring capacity which may not be locally available, and this could increase risks associated with the technology's use."[30] This has been demonstrated where a lack of training in pesticide use has led to food contamination, poisoning, and pesticide resistance. In addition, the lack of consistent regulation, product registration, and effective evaluation are important factors that developing Africa will need to consider as it continues its exploration of these platform technologies.

As a result, it is imperative that developing countries begin to build their own research capacity by upgrading National Agricultural Research Institutes (NARIs) to become universities that have direct linkages with smallholder farmers and agribusinesses. These new universities must examine and evaluate the benefits and risks of new technologies in a local context, and must train farmers to become innovators and entrepreneurs.[31]

Regulation

Regulating biotechnology should take into account evidence from prior global use. Overall, the balance of the evidence tends to support the safety of transgenic crops. For example, the European Commission funded more than 50 studies to evaluate this issue and found that "the use of biotechnology and of GE plants *per se* does not imply higher risks than classical breeding methods or production technologies."[32] A literature review covering the last 10 years of transgenic crop safety and effects on biodiversity and human health concludes that "the scientific research conducted thus far has not detected any significant hazard directly connected with the use of transgenic crops."[33]

Despite this growing body of evidence, developing countries must overcome strong regulatory barriers to adoption of transgenic crops. While transgenic crops have the potential to greatly increase crop and livestock productivity and nutrition, a popular backlash against transgenic foods has created a stringent political atmosphere under which tight regulations are being developed. Much of the inspiration for restrictive regulation comes from the Cartagena Protocol on Biosafety under the United Nations Convention on Biological Diversity.[34] The central doctrine of the Cartagena Protocol is the "precautionary principle" that empowers governments to restrict the release of products into the environment or their consumption even if there is no scientific evidence that they are harmful.

These approaches differ from food safety practices adopted by the World Trade Organization (WTO), which allow governments to restrict products when there is sufficient scientific evidence of harm.[35] Under the Cartagena Protocol, public perceptions are enough to trigger a ban on such products. Those seeking stringent regulation have cited uncertainties such as horizontal transfer of genes from transgenic crops to their wild relatives. Some have raised fears about the safety of transgenic foods to human health. Other concerns include the fear that farmers would be dependent on foreign firms for the supply of seed.[36]

Sub-Saharan Africa in particular follows a strict interpretation of the European regulatory model, which uses the precautionary principle to evaluate transgenic crops (as opposed to the United States, which evaluates the crop itself). Given the differences between US and European regulatory systems, there is a lack of harmonization that hinders the adoption process. Farmers in sub-Saharan Africa also have little political power and cannot make the case for adoption, despite comprising such a large percentage of the population. This is not always the case, however. South Africa, for example, has produced transgenic crops for the past 18 years and has a particularly effective biosafety regulatory framework and R&D investment. South Africa also trained farmers and scientists and embarked on a substantive public awareness campaign. In addition, farmers groups (including both large-scale and smallholder farmers) were supportive of the adoption of transgenic crops.[37]

Because sub-Saharan Africa is tied so heavily to European markets and models, however, policymakers are often swayed by European debates. But the circumstances cannot be compared; Africa faces unique challenges that are absent in Europe. Policymakers in developing countries need to assess the challenges they face and determine appropriate solutions. Many solutions can lead to improved agricultural productivity, higher incomes, and climate change mitigation; these include efficient water use, fertilizer use, better land and soil management practices, improved infrastructure, and more storage facilities. But in certain circumstances, transgenic crops are the best option—such as "when there is only limited genetic variation in the trait of interest in a crop."[38] Conventional breeding failed to produce a *Maruca*-resistant cowpea, a staple crop in many African countries. Another staple, cassava, is affected by two widespread viral diseases. Conventional breeding is not an option because cassava flowers every two years, so Uganda and Kenya are researching whether genetic modification can breed *Bt*-resistant cassava.[39]

Perhaps the most significant barrier to adoption is that sub-Saharan Africa is heavily tied to European markets, which forbid the importation of transgenic crops, except for animal feed. Until other export markets and regional trade options become more viable, farmers do need to be cognizant of market access.

In addition, the cost of implementing these regulations could be beyond the reach of most African countries.[40] Such regulations have extended to African countries, and this tends to conflict with the great need for increased food production.

As rich countries withdraw funding for their own investments in agriculture, international assistance earmarked for agricultural science has diminished.[41]

In Nigeria, the findings of a study on biotechnology awareness demonstrate that while respondents have some awareness of these techniques, this is not the case for biotechnology products.[42] Most of the respondents are favorably disposed to the introduction of transgenic crops and will eat transgenic foods if they are proven to be significantly more nutritious than nontransgenic foods. The risk perception of the respondents suggests that although more people are in favor of the introduction of transgenic crops, they, however, do not consider the current state of Nigeria's institutional preparedness satisfactory for the approval and release of transgenic crops.

However, it is important to consider that African farmers will not grow successful crops if prices are low or dropping. Additionally, complications with regulation and approval of transgenic crops make obtaining commercial licenses to grow certain crops difficult. Also, neighboring countries must often approve similar legislation to cover liabilities that might arise from cross-pollination by wind-blown pollen, for example. Biosafety regulations often stall developments in the research of transgenic crops and could have negative impacts on regional trade.[43]

For these reasons, approval and use of potentially beneficial crops are often difficult. However, despite potential setbacks, biotechnology has the potential to provide both great profits and the means to provide more food to those who need it in Africa. Leaders in the food industry in parts of Africa prefer to consider the matter on a case-by-case basis rather than adopt a generic approach to biosafety.[44] In fact, the tendency in regulation of biotechnology appears to follow more divergent paths reflecting unique national and regional attributes.[45] This is partly because regulatory practices and trends in biotechnology development tend to co-evolve as countries seek a balance between the need to protect the environment and human safety and foster technological advancement.[46]

At the same time, new techniques such as gene editing, described earlier, demand new approaches. As mentioned, genetically edited crops have no foreign genes inserted. Regulatory bodies worldwide are currently grappling with how to classify this new approach and the resulting crops. The precedent for such regulation is one of process (where regulators consider whether biotechnology was used to create a crop) or product (which looks at a new specific attribute). If no foreign genes are present, however, the contrast between US and EU approaches to regulating biotech crops might change.[47] Furthermore, the European Union is now starting to allow member states to make their own decisions regarding transgenic and biotech crops. Ultimately, if Africa decides that these newest biotechnology techniques are in its best interest, countries must establish their own protocols and must decide for themselves how to classify genetically edited crops.

Conclusion

Probably the most significant research and educational opportunities for African countries in biotechnology lie in the potential to join the genomics revolution as the costs of sequencing genomes drop. When James Watson, co-discoverer of the DNA double-helix, had his genome sequenced in 2008 by 454 Life Sciences, the price tag was US$1.5 million. A year later a California-based firm, Applied Biosystems, revealed that it has sequenced the genome of a Nigerian man for under US$60,000. In 2010 another California-based firm, Illumina, announced that it had reduced the cost to about US$20,000. New machines can now sequence a human genome for just $1,000.[48]

Dozens of genomes of agricultural, medical, and environmental importance to Africa have already been sequenced. These include rice, corn, mosquito, chicken, cattle, and dozens of plant, animal, and human pathogens. The challenge facing Africa is building capacity in bioinformatics to understand the location and functions of genes. It is through the annotation of genomes that scientists can understand the role of genes and their potential contributions to agriculture, medicine, environmental management, and other fields. Bioinformatics could do for Africa what computer software did for India. The field would also give African science a new purpose and help to integrate the region into the global knowledge ecology. This opportunity offers Africa another opportunity for technological leapfrogging. The central challenge, therefore, is building capacity in biotechnology research, which includes biosafety.

4

AGRICULTURAL INNOVATION SYSTEMS

The use of emerging technology and indigenous knowledge to promote sustainable agriculture will require adjustments in existing institutions.[1] New approaches will need to be adopted to promote close interactions between government, business, farmers, academia, and civil society. The aim of this chapter is to identify novel agricultural innovation systems of relevance to Africa. It will examine the connections between agricultural innovation and wider economic policies. Agriculture is inherently a place-based activity and so the chapter will outline strategies that reflect local needs and characteristics. Positioning sustainable agriculture as a knowledge-intensive sector will require fundamental reforms in existing learning institutions, especially universities and research institutes. Specifically, key functions such as research, teaching, extension, and commercialization need to be much more closely integrated.

The Concept of Innovation Systems

Agriculture is considered central to African economies, but it is treated like other sectors, each with their own distinctive institutions and with little regard for their relationship with the rest of the economy.[2] This view is reinforced by traditional approaches, which argue that economic transition occurs in stages that involve the transfer of capital from the agricultural to the industrial sector. Both the sector and stage approaches conceal important linkages between agriculture and other sectors of the economy.

A more realistic view is to treat economies as "systems of innovation." The process of technological innovation involves interactions among a wide range of actors in society, who form a system of mutually reinforcing learning activities. These interactions and the associated components constitute dynamic "innovation systems."[3] Innovation systems can be understood by determining what within the institutional mixture is local and what is external. Open systems are needed, in which new actors and institutions are constantly being created, changed, and adapted to suit the dynamics of scientific and technological creation.[4] The concept of a system offers a suitable framework for conveying the notion of parts, their interconnectedness, their interaction, evolution over time, and the emergence of novel structures. Within countries, the innovation system can vary across localities. Local variations in innovation levels, technology adoption and diffusion, and the institutional mix are significant features of all countries.

An innovation system is a "network of organizations, enterprises, and individuals focused on bringing new products, new processes, and new forms of organization into economic use, together with the institutions and policies that affect their behavior and performance. The innovation systems concept embraces not only the science suppliers but the totality and interaction of actors involved in innovation. It extends beyond the creation of knowledge to encompass the factors affecting demand for and use of knowledge in novel and useful ways."[5]

Government, the private sector, universities, and research institutions are important parts of a larger system of knowledge and interactions that allows diverse actors with varied strengths to come together to pursue broad common goals in agricultural innovation. In many African countries, the state still plays a key role in directing productive activities. But the private sector is an increasingly important player in adapting existing knowledge and existing technologies and applying that to new areas.

The innovation systems concept comes from "direct observations of countries and sectors with strong records of innovation. It has been applied to agriculture in developing countries only recently, but it appears to offer exciting opportunities for understanding how a country's agricultural sector can make better use of new knowledge and for designing alternative interventions that go beyond research system investments."[6]

Systems-based approaches to innovation are not new in the agricultural development literature. The study of technological change in agriculture has always been concerned with systems, as illustrated by applications of the national agricultural research system (NARS) and the agricultural knowledge and information system (AKIS) approaches. However, the innovation systems literature is a major departure from the traditional studies of technological change that are often used in NARS- and AKIS-driven research.[7]

The NARS and AKIS approaches, for example, emphasize the role of public-sector research, extension, and educational organizations in generating and disseminating new technologies. Interventions based on these approaches traditionally focused on investing in public organizations to improve the supply of new technologies. A shortcoming of this approach is that the main restriction on the use of technical information is not just supply or availability but also the limited ability of innovative agents to absorb it. Even though technical information may be freely accessible, innovating agents must invest heavily to develop the ability to use the information.

While both the NARS and AKIS frameworks made critical contributions to the study of technological change in agriculture, they are now challenged by the changing and increasingly globalized context in which sub-Saharan African agriculture is evolving. There is a need for a more flexible framework for studying innovation processes in developing-country agriculture—a framework that highlights the complex relationships between old and new actors, the nature of organizational learning processes, and the socioeconomic institutions that influence these relationships and processes.

The agricultural innovation system maps out the key actors and their interactions that enable farmers to obtain access to technologies. The "farm firm" is at the center of the agricultural innovation system framework, and the farmer as the innovator could be made less vulnerable to poverty when the system enables him to access returns from his innovative efforts. The agricultural innovation system framework presents a demand-driven approach to agricultural R&D. This transcends the perception of the role of public research institutions as technology producers and farmers as passive users by viewing the public laboratory-farmer relationships as an interactive process governed by several institutional players that determine the generation and use of agricultural innovation. There is opportunity for a participatory and multi-stakeholders approach to identifying issues for agricultural R&D, and agricultural technology could thus be developed with farmers' active participation and understanding of the application of new technologies. The agricultural innovation system approach as an institutional framework can be fostered depending on the institutional circumstances and historical background of the national agricultural development strategies.[8]

This brings us to the agricultural innovation system (AIS) framework. The AIS framework makes use of individual and collective absorptive capabilities to translate information and knowledge into a useful social or economic activity in agriculture. The framework requires an understanding of how individual and collective capabilities are strengthened, and how these capabilities are applied to agriculture. This suggests the need to focus far less on the supply of information and more on systemic practices and behaviors that affect organizational learning and change. The approach essentially unpacks systemic structures into processes as a means of strengthening their development and evolution.

Recent discussions of innovation capacity have argued that capacity development in many countries involves two sorts of tasks. The first is to create networks of scientific actors around research themes such as biotechnology and networks of rural actors around development themes such as dryland agriculture. The second is to build links between these networks so that research can be used in rural innovation. A tantalizing possibility is that interventions that unite research-led and community-based capacity could cost relatively little, add value to existing investments, meet the needs of the poor, and achieve very high returns.

Innovation Systems in Action

University-Industry Linkages

Universities play a key role in regional innovation systems by providing skilled workers, conducting research adapted to local conditions, and sharing services and technologies with small to medium-sized enterprises and other stakeholders.[9] University involvement in regional innovation systems can take one of two forms. Universities can tangibly respond to regional needs by creating business incubators and science parks. More commonly, however, they alter their teaching, research, and consulting services to meet regional needs.[10]

Trends in university-industry linkages (UILs) in Nigeria illustrate two ways in which university-industry collaboration has been experienced in the Nigerian agro-food-processing sector. They are principal agent demand-driven and multi-stakeholder problem-based. The examples of university-industry interactions in these two modes demonstrate that universities and firms in Nigeria working together can build capacity for innovation. These two modes have contributed to innovative outcomes involving the diffusion and commercialization of local R&D.[11]

A noteworthy example of the first mode of UIL, "principal agent demand-driven," is the UNAAB-Nestlé Soyabean Popularization and Production Project, a successful partnership between the University of Agriculture Abeokuta (UNAAB) and Nestlé Nigeria since 1999. Nestlé employs UNAAB to help address its challenges in the demand for soybeans, which is a major raw material used largely in baby food production, by leveraging the research and extension activities of UNAAB. It is thus plausible to consider the principal agent in this case of UIL as Nestlé, and the driver of the UIL as the demand for soybeans.

The main objectives of the UIL included integrating soybeans into the existing farming systems in the southwestern part of Nigeria; promoting mass production of high-quality grains to meet Nestlé Nigeria's standards; and improving the welfare of the farmers by stimulating their interest in sustainable soybean production.

The UIL can be initially traced to an R&D partnership under a tripartite agreement for soybean breeding between UNAAB, the International Institute of Tropical Agriculture (IITA), Ibadan, and Nestlé Nigeria in the early 1990s. Nestlé Nigeria financed the soybean-breeding project to produce high-quality soybeans with significantly improved yields. Although the research team achieved this initial objective in 1996, with the breeding of Soya 1448–2E, the partnership over time has also led to the popularization of soybeans in southwest Nigeria, after UNAAB's research established that soybeans can also be grown outside Nestlé's original geographic focus of northern Nigeria.

There are a number of benefits for such university-industry linkages: farmers contributed significantly to building capacity for innovation, especially at the farm level; there were huge improvements in the quality of seeds and grains; and a new process for growing soybeans was developed. Nestlé Nigeria saved costs by finding alternatives to the inefficient Nestlé Nigeria farms located in northern Nigeria and secured a regular supply of high-quality soybeans from farmers in the UIL. The system boosted UNAAB's extension activities, resulting in the popularization of its model of soybean cultivation in southwest Nigeria, which in turn became an important soybean-producing region. Overall, the linkages improved the livelihoods of the people in the region and enhanced technology adoption for soybean processing, especially threshing technology.

The second mode of UIL, identified as "multi-stakeholder problem-based," is the Cassava Flash Dryer Project. The project involved a large, privately owned, integrated farm (Godilogo Farm, Ltd.) with an extensive cassava plantation and processing factory; three universities including the University of Agriculture, Abeokuta, the University of Ibadan, and the University of Port Harcourt; the IITA; and the Raw Material Research and Development Council (RMRDC).

Cassava is Africa's second most important food staple, after maize, in terms of calories consumed, with potential to address the challenges of food security and welfare improvement. Nigeria is currently the world's largest producer of cassava. The Presidential Initiative on Cassava Production and Export (PICPE) was officially launched in 2004 to promote industrial processing and exporting of cassava products. Support for research on cassava processing and cassava products allowed stakeholders to address the challenge of cassava production and industrial processing, including the design and fabrication of a cassava flash dryer.

Though the principle of flash drying is well known in engineering theory and practice, it is not widely applied to indigenous agricultural crops in Nigeria, partially due to a design gap in understanding the engineering properties of most of the Nigerian crops. The flash dryers available in the market are designed for agricultural products that are grown in industrialized countries such as Irish potatoes or maize. They are usually modified with the help of foreign technical partners for use in cassava processing, which often results in lower performance and frequent equipment breakdowns. This was the experience of Godilogo Farms, Ltd., which had used a flash dryer imported from Brazil, because the design was unable to handle drying the cassava to the required moisture or water content. The main objective of the Cassava Flash Dryer Project was to design and fabricate an efficient cassava flash dryer to withstand the stress of the local operating environment.

The new locally produced cassava flash dryer designed by the PICPE-IITA research team produces 250 kilograms of cassava flour per hour. The RMRDC funded the official commissioning of the new flash dryer at Godilogo Farms, Obudu, Cross Rivers State, on August 19, 2008. IITA and PICPE provided the initial funding under the IITA Integrated Cassava Project; the Root and Tuber Extension Program supported the design team's visit to collect data from existing flash drying centers; Godilogo paid for the fabrication of the plant and part sponsorship of the researchers' living costs; and RMRDC provided logistical support for several trips by the design team, including sponsorship of the commission.

The technological and interactive learning through experimentation generated from the creation of the first medium-sized cassava flash dryer was unprecedented in the local fabrication of agro-food-processing equipment. The impact of government policy through PICPE and government support for the project through RMRDC demonstrated the crucial role of government as a mediator or catalyst for UIL and innovation. Knowledge flows and user feedbacks also played important roles in the success of the university-industry linkage.

In addition to the Nigerian projects outlined above, a shortage of highly skilled personnel combined with increased demand from commercial producers within the agricultural sector has led to multiparty ventures between training institutions, agricultural industry groups, and third-party donors. This has been evident in high-value agricultural export industries such as floriculture and horticulture in Uganda and Ethiopia,[12] where farmers were previously hiring foreigners for middle management positions given the lack of local candidates possessing the necessary skill set.

In particular, the Ugandan Flower Exporters Association and the Ethiopian Horticulture Producers and Exporters Association felt the need to train local talent. These farmers and producers enlisted the Netherlands Foundation for International Cooperation to develop a variety of training options, ranging from short courses for farm workers to certificate and diploma courses to bachelor and master's programs. Crucial to the success of this initiative was liaising with producers to identify the necessary knowledge and skills required to fill the supply and demand gap, in addition to developing a new pedagogical, hands-on teaching approach. The project resulted in competence-based training, ensuring that trained technicians and middle managers were equipped with the appropriate work-ready skills. By 2012, within the first batch of 16 diploma students in Uganda, 14 were working on local flower farms.

This example highlights that employers can and should drive the demand for vocational training. When this is combined with close collaboration from a training supplier, the results can be hugely beneficial to both the individual receiving the training and the productive sector client.

Certain universities across Africa have taken the initial steps to establishing linkages within the agricultural industry to further increase productivity. However, studies report that many universities have minimal linkages with the productive sector, ranging from agricultural producers to big industry and small and medium-sized enterprises.[13] It is clear that the benefits of UILs are not being fully taken advantage of and that there is still potential for such initiatives to be scaled up.

Wider Institutional Linkages

Understanding the network relationships and institutional mechanisms that affect the generation and use of innovation in the traditional sector is critical for enhancing the welfare of the poor and overall economic development. Nigeria's development policy emphasizes making agriculture and industrial production the engine of growth. In recent years the revitalization of the cocoa industry through the cocoa rebirth initiative launched in February 2005 has been a major focus of government.[14]

The program essentially aimed at generating awareness of the wealth creation potentials of cocoa, promoting increases in production and industrial processing, attracting youth into cocoa cultivation, and helping to raise funds for the development of the industry. By applying the analytical framework of the agricultural system of innovation, it is easier to trace the process of value-addition in the cocoa agro-industrial system, examining the impact of the cocoa rebirth initiative and identifying the actors critical for strengthening the cocoa innovation system in Nigeria.

Cocoa production is a major agricultural activity in Nigeria; and R&D aimed at improving cocoa production and value-addition has long existed at the Cocoa Research Institute of Nigeria (CRIN) and notable faculties of agriculture in Nigerian universities and colleges of agriculture. However, while the export of raw cocoa beans has continued to thrive, innovation in cocoa production and the industrial processing of cocoa into intermediate and consumer products have been limited.

The cocoa innovation system in Nigeria is still relatively weak. There is a role for policy intervention in stimulating interaction among critical agents in this agricultural innovation system. In particular, linkages and interactions between four critical actors (individual cocoa farmers, cocoa-processing firms, CRIN, and the National Cocoa Development Committee) in the cocoa rebirth program were identified as being responsible for the widespread adoption of CRIN's newly developed genetically improved cocoa seedlings, capable of a yield exceeding 1.8 metric tonnes per hectare per year. This is in stark contrast to the previous experiences of CRIN, which has been unable to commercialize many of its research findings. Periodic joint review of the activities of each of these actors and active participation in specific projects that are of common interest may further innovation, especially in value-addition to cocoa beans.

The adoption and diffusion of improved cocoa seedlings under the cocoa rebirth initiative thrive on subsidies provided by government. While subsidies for agricultural production in a developing country such as Nigeria may not be discouraged, it is important to have a phased program of subsidy withdrawal on the cocoa seedlings program when it is certain that farmers have proven the viability and economic importance of the new variety. This should result in a market-driven diffusion that will be healthy for the sustainable growth of the cocoa industry.

Despite success with the diffusion of cocoa seedlings, the findings show that although export is a major concern of the cocoa-processing firms, and this appeared to have led to close interactions of the firms with the National Export Promotion Council (NEPC), the export strategy has not been effectively linked with the cocoa rebirth initiative. In order to further encourage export by the cocoa-processing firms, it would be good to integrate the NEPC export incentives into the cocoa rebirth initiative within the cocoa innovation system framework. Moreover, the NEPC should also adopt an innovation system approach to export strategy. This would essentially begin by emphasizing demonstrable innovative activities of firms as an important requirement for the firms to benefit from export incentives.

The involvement of the financial sector in the cocoa innovation system is identified as a main challenge. Though the financial sector is aware of the significance of innovation for a competitive economy, its response to the cocoa rebirth initiative has been slow due to perceived relatively low return on investments. It is suggested that the publicly owned Bank of Industry and the Central Bank of Nigeria (CBN) should provide leadership in investing in innovative new start-ups in cocoa processing and in carefully identified innovative ideas or projects in existing cocoa-processing firms. This demonstration should be carried out in partnership with interested commercial banks with the CBN, guaranteeing the banks' investment in the project. Once the banks are convinced that innovative initiatives in firms are able to provide satisfactory returns on investment, they should be open to investing in such projects.

Skills deficiency is a major constraint on the cocoa innovation system. The result suggests that skills development in the areas of cocoa farm management and the operation of modern cocoa-processing machinery would be particularly useful in enhancing cocoa output and the performance of cocoa-processing firms. In this respect, renewed efforts are needed by the educational and training institutions to improve the quality and quantity of skills being produced for cocoa-processing firms.

As part of the cocoa rebirth initiative, special training programs should be organized for skills upgrading and new skills development relevant to the cocoa industry. Another important constraint on the cocoa innovation system arises from the difficulty in implementing the demand-side aspects of the cocoa rebirth initiative, such as serving free cocoa beverages in primary schools and using cocoa-based beverages in government offices, practices that should stimulate innovative approaches to increased local processing of cocoa and the manufacture of cocoa-based products.

Clusters as Local Innovation Systems

Theory, evidence, and practice confirm that clusters are important source of innovation.[15] Africa is placing considerable emphasis on the life sciences. There is growing evidence that innovation in the life sciences has a propensity to cluster around key institutions such as universities, hospitals, and venture capital firms.[16] This logic could be extended to thinking about other opportunities for clustering that include agricultural regions. Essentially, clusters are geographic concentrations of interconnected companies and institutions in a particular field. Clusters encompass an array of linked industries and other entities important to competition. They include, for example, suppliers of specialized inputs such as components, machinery, and services, and providers of specialized infrastructure.

The existing literature suggests that there are three types of clusters in Africa: (1) the groundwork cluster, which improves the producers' access to markets; (2) the industrializing cluster, which starts the process of specialization and differentiation; and (3) the complex cluster, which has already diversified and can begin accessing wider national and international markets.[17] African clusters are in an early stage of development, meaning that most initiatives are contained within the groundwork and industrializing cluster types.

The lack of complex clusters in an African context is partially due to the following structural barriers that should be addressed to further propel innovation and growth: underdeveloped regional trading networks in Africa; weak political and economic institutions; cluster occurrence in areas with an overabundance of labor, resulting in less effective labor-pooling initiatives; and premature market liberalization of large-scale industries, rendering it more difficult for small and medium-sized firms to compete with an abundance of imports.

Often clusters extend downstream to channels and customers and laterally to manufacturers of complementary products, as well as to companies in related industries, either by skills, technologies, or common inputs. Finally, many clusters include governmental and other institutions—such as universities, standard-setting agencies, think tanks, vocational training providers, and trade associations—that provide specialized training, education, information, research, and technical support.[18] The co-evolution of all actors supports the development of dynamic innovation systems, which accelerate and increase the efficiency of knowledge transfer into products, services, and processes and promote growth. As clusters enable the flow of knowledge and information between enterprises and institutions through networking, they form a dynamic self-teaching system and they speed up innovation. Local knowledge develops that responds to local needs—something that rivals find hard to imitate.

Although much of the recent literature on clusters focuses on small to medium-sized high-tech enterprises in advanced industrial countries, a smaller school of literature has already begun expanding the study of clusters to include agricultural innovation. Clusters can and often do emerge anywhere that the correct resources and services exist. However, central to the idea of clusters is the concept that positive "knowledge spillovers" are more likely to occur between groups and individuals that share spatial proximity, language, culture, and other key factors usually tied to geography.

Contrary to scholars who argue that the Internet and other information technologies have erased most barriers to knowledge transfer, proponents of cluster theory argue that geography continues to dominate knowledge development and transfer, and that governments seeking to spark innovation in key sectors (including agriculture) should therefore consider how to encourage the formation and growth of relevant clusters. A key intuition in this argument is that informal social interactions and institutions play a central role in building trust and interpersonal relationships, which in turn increase the speed and frequency of knowledge, resource, and other input sharing.

In developing countries, clusters are present in a wide range of sectors and their growth experiences vary widely, from being stagnant and lacking competitiveness to being dynamic and competitive. This supports the view that the presence of a cluster does not automatically lead to positive external effects. There is therefore a need to look beyond the simple explanation of proximity and cultural factors, and to ask why some clusters prosper and what specifically explains their success.

Shouguang Vegetable Cluster, China

China has a long history of economic clusters in sectors as diverse as silk, porcelain, high technology, and agriculture.[19] One of China's most successful agricultural clusters is the vegetable cluster in Shandong Province. This "Vegetable City" is a leading vegetable production, trading, and export center. Its 53,000-hectare vegetation plantation produces about four million tons annually. Shouguang was one of the poorest areas in the Shandong province until the early 1980s, when vegetable production started. Today five state- and provincial-level agricultural demonstration gardens and 21 nonpolluted vegetable facilities have been established. More than 700 new vegetable varieties have been introduced from over 30 countries and regions. Shouguang also hosts China's largest vegetable seed facility aimed at developing new varieties. The facility is co-sponsored by the China Agricultural University. Over the years, vegetable production has increased, leading to the emergence of an agro-industrial cluster that has helped to raise per capita income for Shouguang's previously impoverished rural poor.[20] The cluster evolved through four distinctive phases.

In the first emergence phase (1978–1984), Shouguang authorities launched programs for massive vegetable planting as a priority for the local development agenda. Shouguang had three main advantages that helped it to emerge as a leading vegetable cluster. These included a long history and tradition of vegetable production, rising domestic and international demand for vegetables, and higher profits exceeding revenue from crops such as rice and wheat. In 1983 Shouguang's vegetable production exceeded 450 tonnes. The local market could not absorb it all, so about 50 tonnes went to waste. The loss prompted Shouguang to construct a vegetable market the following year, thereby laying the foundation for the next phase.

In the second phase of the development of the cluster, local government officials used their authority to bring more peasants and clients into the new market. For example, the officials persuaded the Shengli Oil Field, China's second largest oil base, to become a customer of Shouguang vegetables. This procurement arrangement contributed to the market's early growth. The authorities also helped to set up more than 10 small agricultural product markets around the central wholesale market, creating a market network in the city. The markets directly benefited thousands of local farmers. Despite these developments, high demand for fresh vegetables in winter exceeded the supply.

The third phase of the development of the cluster was associated with rapid technological improvements in greenhouses and increased production. In 1989, Wang Leyi, chief of a village in Shouguang, developed a vegetable greenhouse for planting in the winter, characterized by low cost, low pollution, and high productivity. This inspired local farmers to adopt the technology and led to incremental improvements in the construction and maintenance of greenhouses. Communication among farmers and the presence of local innovators helped to spread the new technology. By the end of 1996, Shouguang had 210,000 greenhouses, and the vegetable yield had grown to 2.3 million tonnes. The Shouguang government focused on promoting food markets. It helped to create more than 30 large specialized markets and 40 large food-processing enterprises. In 1995 the central government authorized the creation of the "Green Channel," an arrangement for transporting vegetables from Shouguang to the capital, Beijing. The transportation and marketing network evolved to include the "Green Channel," the "Blue Channel" (ocean shipping), the "Sky Channel" (air transportation), and the "Internet Channel."

After 1997 the cluster entered its fourth development phase, which involved the establishment of international brand names.

The internationalization was prompted by the saturation of domestic markets and rising nontariff trade barriers, such as strict and rigorous standards. International safety standards and consumer interest in "green products" prompted Shouguang to establish 21 pollution-free production bases. Foreign firms such as the Swiss-based Syngenta Corporation played a key role in upgrading planting technologies, providing new seed and offering training to local farmers. This was done through the Shouguang Syngenta Seeds Company, a joint venture between Syngenta and the local government. Syngenta signed an agreement with the Ministry of Agriculture's National Agricultural Technical Extension and Service Center to train farmers in modern techniques. Since 2000, the one-month (starting April 20) Shouguang vegetable fair has encapsulated and perpetuated this cluster's many successes and has become one of China's premier science and technology events.

Rice Cluster in Benin

Entrepreneurship can spur innovations, steer innovation processes, and compel the creation of an innovation-enabling environment while giving rise to and sustaining the innovation system. Entrepreneurial venture is an embedded power that steers institutions, stimulates learning, and creates or strengthens linkages that constitute the pillars of innovation systems. The dissemination of New Rices for Africa (NERICA) in Benin illustrates what can be considered a "self-organizing innovation system."[21] This section describes NERICA's unique approach, combining the innovation systems approach and entrepreneurship theory, which enabled a class of entrepreneurs to take the lead in the innovation process while creating the basis for a system of innovation to emerge.

Benin, which is located in West Africa, covers an area of 112,622 square kilometers and has nearly 8.2 million inhabitants. Its landscape consists mostly of flat to undulating plains but also includes some hills and low mountains. Agriculture is the predominant basis of the country's weak economy; although only contributing 32% of the GDP (as compared to 53.5% of the service sector and 13.7% of the industrial sector), it employs about 65% of the active population.

Despite relatively favorable production environments, Benin's domestic production is weak and meets only 10%–15% of the country's demand for rice. Different people attribute this to different causes, such as policies and institutions that are not suited to supporting domestic production against importations or low quality of products. Irrigation possibilities are not fully exploited, despite the fact that rice production is traditionally rain-fed. There is also minimum input, with improper seeding and lack of fertilizers, pesticides, and herbicides.

NERICA is the brand name of a family of improved rice varieties specially adapted to the agro-ecological conditions of Africa. It is a hybrid that combines the best traits of two rice species: the African *Oryza glaberrima* and the Asian *Oryza sativa*. It has certain advantages over other species, such as high yields, quick maturity, and resistance to local biotic and abiotic stresses such as droughts and iron toxicity. It also has 25% higher protein content than international standard varieties. And it is more responsive to fertilizers. Due to these advantages, different groups that wanted to change the status quo of Benin's agriculture sought to introduce NERICA. They included the government of Benin, the Banque Régionale de Solidarité (BRS), agro-industrial firms such as Tunde Group and BSS-Société Industrielle pour la Production du Riz (BSS-SIPRi), as well as nongovernmental entities such as Songhaï, Projet d'Appui au Développement Rural de l'Ouémé (PADRO), and Vredeseilanden (VECO). These organizations worked closely together to bring to the task skills, knowledge, and interests that could not be found in one entity.

A simple introduction of all of these organizations helps to clarify how they converged on NERICA in their pursuit of agricultural innovation. Songhaï is a socioeconomic and rural development NGO specializing in agricultural production, training, and research. It supports an integrated production system that promotes minimal inputs and the use of local resources. Songhaï was one of the first pioneers of NERICA production in Benin, largely because it was challenged to endorse a framework conducive to rice production as a profitable commodity.

Songhaï came in contact with BRS as it was seeking to fund skilled, competent, and innovative economic agents with sound business plans. Songhaï fit the bill perfectly. Tunde Group was NERICA's production hub and BSS-SIPRi is an enterprise specializing in NERICA seed and paddy production. PADRO and VECO are NGOs from France and Belgium, respectively. PADRO worked with the extension agency, farmer organizations, and micro-finance establishments, and indirectly with the Ministry of Agriculture. VECO focused on culture, communication, sustainable agriculture, and food security.

All of these separate organizations came together through NERICA to challenge Benin's agricultural status quo. Their entrepreneurism not only directly helped the dissemination of NERICA but also pushed the Benin government toward policies for agricultural business development. In February 2008, the government issued a new agricultural development strategy aiming to establish an institutional, legal, regulatory, and administrative environment conducive to agricultural activities.

What can be learned from the NERICA case is that the dissemination of this new technology did not follow the conventional process of assistance programs and government adoption. There was a process of self-organization through various nongovernmental organizations. Self-motivated economic entrepreneurs started the process and propelled innovation. As a result, the private sector was able to push the government to adopt new policies that would be conducive to these innovations. These conditions then created more economic opportunities, which drew more self-organized entrepreneurs to the program and thereby completed a healthy cycle of economic and technological improvement. This process as a whole can be understood as a self-organizing system of innovation.

Wine Cluster in South Africa

University-industry linkages are most valuable when it comes to knowledge diffusion, which positively impacts the economy. In the wine industry specifically, university research played an important role in the phylloxera outbreak in the 1860s. Since then, university scientists carefully researched innovative and practical agronomic, chemical, and engineering solutions to other industry problems. Researchers offered advice to farmers and industry professionals in recognizing and treating pests and viral pathogens, analyzing soil components, and implementing innovative irrigation options in the face of climate change.[22]

South Africa boasts a vibrant wine industry that can be traced back to the Dutch settlers in the seventeenth century. Starting in the 1980s, with deregulation of the industry, South Africa began focusing on quality as well as on taking advantage of production technologies and growing techniques to compete successfully with traditional winemakers such as France, Italy, and Spain. Much of the success of the wines, which are now a large part of the tourism and export sectors of the South African economy, can be attributed to the well-established wine cluster, particularly around the Western Cape province. The end of Apartheid in 1994 and the abolishment of the quota system provided the impetus for start-ups to enter the industry and small wineries to expand. South Africa is now the ninth largest wine producer in the world, and within the New World wine countries has a production share of 9.0% and an export share of 13.7%, with the United Kingdom and the Netherlands representing the majority of the export demand.[23]

It can be argued that the success of the South Africa wine cluster is largely due to well-established collective actions and strong institutional support.[24] The first instance of cooperative action occurred with the formation of the Cooperative Viniculture Organization in 1917, which organized the industry into cooperative producers and growers in addition to establishing quality and price controls to re-stabilize the industry following the Anglo-Boer War. Recognizing its limitations in breaking into the global market, however, the industry established the South African Wine and Brandy Company, which is a cluster representing all stakeholder interests. The company focuses on research and development, marketing, and technical expertise. Winetech, a subset of this company, coordinates all research activities in the industry. Today, exchanges between key producers and sellers within the industry help foster innovation at the technical and organizational level, which has translated into transformational growth. The benefits range from support service contractors who are expanding export wines in premium segments to joint marketing efforts between producers and institutions.

In addition, a few public and nonprofit institutions assist the South African wine industry. The most prominent of these is the state-funded Nietvoorbij Institute for Viticulture and Oenology, which is part of the Agriculture Research Council. The cluster also benefits from strong university-industry linkages with the departments of viniculture and viticulture at the University of Stellenbosch in the Western Cape to provide academic and research support. These two research institutions account for 90% of the research in the industry; the projects focus on industry needs, and they often provide training to the end users. Finally, an independent nonprofit, Wines of South Africa, has been responsible for the international promotion of South African wines since 2000, and the South African Wine Information Service assists with the collection, processing, and dissemination of industry information. Today, South Africa's wine cluster and its innovation system are robust, and it has some of the strongest university-industry linkages as a result of Winetech, which is unique among wine-producing countries.[25]

Clusters provide crucial formal and informal linkages that increase trust among diverse actors, leading to greater exchange of individuals and ideas and cooperation in key areas that no single firm or institution could achieve on its own. Despite advances in telecommunications, innovation in many sectors continues to be generated by and most easily transmitted between geographically proximate actors.

As farmer productivity is often constrained by lack of appropriate technology or access to best practice knowledge, inputs, and services, clusters may be able to provide pronounced benefits in the agro-sector. Certain types of clusters may have a more direct impact on poverty. These are the clusters in rural areas and in the urban informal economy; clusters that have a preponderance of SMEs, micro-enterprises, and home workers; clusters in labor-intensive sectors in which barriers to entry for new firms and new workers are low; and clusters that employ women, migrants, and unskilled labor.

In many African countries the agricultural sector is dominated by family-based small-scale planting. This structure slows down the diffusion and adoption of information and modern technology, a key driver of agricultural productivity and net growth. One of the main challenges is therefore to enhance technology transfer from knowledge producers to users in the rural regions where small-scale household farming dominates. Clusters can overcome these shortcomings by creating the linkages and social capital needed to foster innovation and technology transfer. However, clusters are not a cure-all for African agricultural innovation, and we must therefore look closely at the conditions under which clusters can work, the common stages of their development, and key factors of their success.

Clusters cannot be imposed on any landscape. They are most likely to form independently or to succeed once seeded by government when they are collocated with key inputs, services, assets, and actors. Clusters are most likely to form and succeed in regions that already possess the proper input, as well as in industries that have a dividable production process and a final product that can be easily transported. Clusters are also more likely in knowledge- or technology-intensive businesses (like agriculture), where breakthroughs can instigate quick and significant increases in productivity. Clusters also benefit from preexisting tightly knit social networks, which provide fertile ground for more complex knowledge generation and sharing infrastructure.

Policies for Cluster Development

Cluster development could benefit from the experiences outlined in the preceding sections. However, despite these examples, cluster work in Africa is at a preliminary stage, which leaves much room for intervention at the policy and program level. In the first phase, governments should lead the formation of clusters by identifying strategic regions with the right human, natural, and institutional resources to establish a competitive advantage in a key sector. Governments can then nurture a quick flow of investment, ideas, and even personnel from the public sector to private firms. As government-funded initiatives deliver proof of concept, governments should make way for private enterprise and give up their ownership stakes in the burgeoning agro-industries they helped create.

As government involvement decreases, clusters move to formalize the connections between key actors through producer associations and other cooperative organizations. Strong bonds formed in the early phases of cluster formation allow diverse actors to come together on common sets of standards in key areas of health, safety, and environment. Quality control and enhanced production are critical for clusters to move beyond their local markets and into more lucrative national or international export markets. Despite their decreasing role, governments can continue to play a key part in this process by putting in place regulations that ease, rather than obstruct, firms' efforts to meet complex international health, environmental, and labor standards.

This strong foundation in place, clusters can move to additional cooperative efforts focused on international marketing and export, and complex partnerships with large multinational companies. Firms can band together to accomplish what none of them can do individually: achieve national and international brand recognition.

Innovation systems likewise cannot be imposed by outside actors and must have substantial buy-in from local government, business groups, and citizen groups. Additionally, governments must wrestle with the possibility that although clusters enhance knowledge generation and transmission within themselves, strong social and practical connections within clusters may actually make communications between them less likely.[26] Linkages between clusters are therefore critical, and this is an area in which regional organizations can play a particularly important role.

Local governments played a critical role in determining initial potential for clustering by evaluating natural and human resources, already existing clusters, and markets in which their area might be able to deliver a competitive advantage. Local governments also assessed and in many cases fueled popular citizen, business, and public institutional support for enhanced cooperation, a key precursor for clusters. As clusters depend on physical and cultural proximity to encourage knowledge creation and sharing, local governments can encourage these exchanges between firms, individual producers, NGOs, and research and academic institutions, even before funding has been set aside for a specific cluster.

While local authorities are best placed to determine the potential for clusters in specific areas, national governments may be better positioned (particularly in Africa) to provide the financial and regulatory support necessary for successful clusters. National governments use state-owned banks, tax laws, and banking regulations to encourage loans to businesses and organizations in these key clusters. They also help finance investments by constructing key infrastructure, including ports, roads, and telecommunications. Finally, governments play a key role in responding to pressure from the clusters to create regulatory frameworks that help them to meet stringent international environmental, health, and labor standards. National governments can also play a central role in convincing nationally funded research and academic institutions to participate actively in clusters with businesses and individual producers.

While clusters lower barriers to knowledge creation and sharing within themselves, the opposite may be true across different national or regional economic activities. This isolation may limit innovation within clusters—or worse, could lead to negative feedback cycles based on the phenomenon of "lock in," whereby clusters increasingly focus on outdated or noncompetitive sectors or strategies.[27] Regional institutions and linkages can play a key role in making and maintaining these external links by supporting the exchange of information, and in particular personnel, between clusters. In Africa, regional institutions could also support the idea of regional centers of excellence based around key specialties—for example, livestock in East Africa.

The Role of Local Knowledge

Strengthening local innovation systems or clusters will need to take into account local knowledge, especially given emerging concerns over climate change.[28] Farming communities have existed for a millennium; long before there were modern agricultural innovations, these communities had to have ways to manage their limited resources and keep the community functioning. Communities developed local leadership structures to encourage participation and the ideal use of what limited resources were available. In the past few centuries, colonial intervention and the push for modern methods have often caused these structures to fail as a result of neglect or active destruction. However, these traditional organizational mechanisms can be an important way to reach a community and cause its members to use innovations or sustainable farming techniques.[29]

While governments and international organizations often overlook the importance of traditional community structures, they can be a powerful tool to encourage community members in the use of new technologies or the revival of traditional methods that are now recognized as more effective.[30] Communities retain the knowledge of and respect for these traditional leadership roles and positions in a way that outside actors cannot, and they will often adopt them as a way to manage community agricultural practices and learning. It is this place-based innovation in governance that accounts to a large extent for institutional diversity.[31]

India's recent reintroduction of the Vayalagams as a means of water management serves as a good example of how traditional systems can still serve the local communities in which they originated as a means of agricultural development and economic sustainability. A long-standing tradition in India in the pre-colonial period was the use of village governance structures called Vayalagams to organize and maintain the use of village water tanks. These tanks were an important component of rain-fed agriculture systems and provided a reservoir that helped mitigate the effects of flooding and sustain agriculture and drinking-water needs throughout the dry season by capturing rainwater.

The Vayalagams were groups of community leaders who managed the distribution of water resources to maximize resources and sustainability, and to ensure that the whole community participated in, and benefited from, the appropriate maintenance of the tanks. Under British colonial rule, and later under the independent Indian government, irrigation systems became centralized, and communities were no longer encouraged to use the tanks, so both the physical structures and the organizations that managed them fell into disrepair.

As the tank-fed systems fell apart and agricultural systems changed, rural communities began to suffer from the lack of sufficient water to grow crops. One solution to this problem has been to revitalize the Vayalagam system and to encourage the traditional community networks to rebuild the system of tanks. Adopting traditional methods of community organization has tapped into familiar resources and has allowed the Development of Humane Action (DAHN) Foundation—an Indian NGO—to rally community ownership of the project and thus gain support for rebuilding the system of community-owned and managed water tanks. The tanks were a defunct system when the DHAN Foundation incorporated in 2002. Now, the Tank-Fed Vayalagam Agricultural Development Programme works in 34 communities and has implemented 1,807 microfinance groups that comprise 102,266 members. The program is funded by a 50% community contribution, with the rest of the funding coming from the foundation. This redeployment of old community organizations has resulted in rapid proliferation of ideas and recruitment of farmers.

Reforming Innovation Systems

As African countries seek to promote innovation regionally, they will be forced to introduce far-reaching reforms in their innovation systems to achieve two important goals. The first will be to rationalize their research activities in line with the goals of the Regional Economic Communities. The second will be to ensure that research results have an impact on the agricultural productive sector. Many emerging economies have gone through such reform processes. China's reform of its innovation system might offer some insights into the challenges that lie ahead.

Partnerships between research institutes (universities or otherwise) and industries are crucial to encourage increased research and promote innovation. Recent efforts in China demonstrate the importance of "motivating universities and research institutes (URIs), building up the innovative capacities of enterprises, and promoting URI-industry linkages."[32] Science and technology (S&T) reforms allowed for increased flexibility, providing incentives to research institutes, universities, and business enterprises to engage in research leading to patents, publications, and other innovations.

During the pre-reform period from 1949 to the 1980s, China focused on a centralized military research model similar to the former Soviet Union, carried out for the most part by public research institutes. Almost all research was planned and funded by the government with individual enterprises (which often had their own S&T institutes and organizations) engaging in little to no research and development.

With the hope of developing the country through education and research, China created the slogan "Building the nation through science and education" to underscore their 1985 reforms. Efforts were made to increase university and research institution collaboration with related business industries.

Reforms occurred in three stages, the first of which spanned 1985–1992. Here, the government encouraged universities and research institutes (URIs) to bolster their connections with industry—one method used was to steeply cut the research budget for universities and other institutes, causing the URIs to turn to industry for support and thus facilitating linkages and partnerships. By the end of 1992, 52 high-tech development zones had been set up, with 9,687 enterprises and a total turnover of renminbi (RMB) 56.3 billion.

From 1992 to 1999, the second stage of reform saw the creation of the S&T Progress Law and the Climbing Program to encourage research as well as the increased autonomy regarding research given to URIs, following the endorsement of enterprises that were affiliated with URIs in 1991. Linkages that were encouraged included technical services, partnerships in development, production, and management, as well as investment in technology. From 1997 to 2000, university-affiliated enterprises experienced average annual sales income growth of 32.3%, with 2,097 high-tech ones emerging in China with a total net worth of US$3.8 billion by 2000.

During the third stage, starting in 1999, China sought to both strengthen the national innovation systems and facilitate the commercialization of R&D results. One key measure was the transformation of state-owned applied research institutes into high-tech firms or technical service firms: a total of 1,149 transformations were carried out by the end of 2003.

New policies and programs helped bring about changes during the reform period. The Resolution on the Reform of the S&T System, released in 1985, aimed to improve overall R&D system management, including encouraging research personnel mobility and the integration of science and technology into the economy through the introduction of flexible operating systems. Peer review of projects and performance brought about a degree of transparency.

One particular program was extremely important in the high-tech area. The 863 Program, which was launched in 1986, sought to move the country's overall R&D capacity to cutting-edge frontiers in priority areas such as biotechnology, information, automation, energy, advanced materials, marine, space, laser, and ocean technology. The 863 Program also promoted the education and training of professionals for the twenty-first century by mobilizing more than 10,000 researchers for 2,860 projects every year. Another example was the Torch Program, launched in 1988. By reducing regulation and building support facilities, "53 national high-tech zones had been established" from 1991–2003, especially in the information technology, biotechnology, new materials, and new energy technologies industries. "The national high-tech zones received RMB 155 billion investments in infrastructure and hosted 32,857 companies in 2003."[33] It appears that these early but critical reform efforts have put China on a path that could enable it to catch up with the industrialized countries in science and innovation.[34]

Because the ultimate goal of the science and technology reforms in China were meant to strengthen national innovation systems and promote innovation activities among the key players in the system, it was necessary for URIs, industry, and the government to interact. The impact of the reforms is seen in the stark contrast between the years 1987 and 2003. In 1987, government-funded public research institutes dominated R&D research, with universities carrying out education and enterprises involved in restricted innovations in "production and prototyping," so URIs found no reason to conduct applied research or to commercialize their research results.

By 2003, R&D expenditure had risen by more than eightfold. Most distinctive was the large increase in R&D units, employees, and expenditures of enterprises. This was brought about in part from the transformation of about 1,000 public research institutes into enterprises or parts of enterprises. Additionally, after the 1991 endorsement of university-affiliated enterprises, a great expansion occurred such that by 2004, 4,593 of them existed, with an annual income of RMB 97 billion.

Another success factor of S&T reforms was increased competition, which created incentives to engage in R&D. This is also apparent in the improved URI-industry linkages, as is shown in the decrease in government spending from 79% in 1985 to 29.9% in 2000. URIs (either transformed or public ones) have forged close links with the private sector "through informal consulting by university researchers to industry, technology service contracts, joint research projects, science parks, patent licensing, and URI-affiliated enterprise."[35] Another success from the S&T reform is the great increase in patents from domestic entities as well as the larger number of publications.

Despite the great success of linkages between industry and institutes for research and education, there are a few cautionary lessons to be learned from China's actions. For example, there has been a lack of focus on science and technology administration. Because the many governmental and nongovernmental bodies work independently, there is a danger of inefficiency in the form of redundancy or misallocation of R&D resources. The reform's focus on commercializing S&T has also prevented further development of basic research and other research aimed at public benefit (with such research stuck at 6% of all research funding). A final concern is the controversy surrounding university-affiliated enterprises that emphasize the operation, ownership structure, and the delinking of such enterprises from their original parent universities. Critics believe that commercial goals may hinder other university mandates about pure academics. When creating comprehensive reform of such magnitude, one must be careful to take into account these potential issues.

China's science and technology reforms demonstrate the potential for expanding research by supporting the formation of URI-industry partnerships and linkages. The benefits are clear, and developing countries should greatly consider using China's case as a model for the establishment of similar programs and policies.

African countries can also draw from the Brazilian agricultural sector's successful agricultural innovation and transformation from a traditional system with low technological usage to a global agricultural pioneer. Between 1985 and 2006, total agricultural production grew by 77%, largely due to effective public investments in science and technology, combined with an environment of economic liberalization and stability.[36] In particular, the rapid modernization of agriculture observed in the 1970s and early 1980s was largely a result of "coordinated policies that led to increased R&D capacity and increased volumes of credit, tied to support policies of stock management, improved distribution and commercialization of food and agro industrial products."[37]

During the agricultural reforms of the last few decades, the Brazilian Agricultural Research Corporation (EMBRAPA), an agricultural research agency funded by the Ministry of Agriculture and Food Supply, was instrumental in boosting Brazilian total factor productivity growth. The agency was initially responsible for providing extension services for the distribution of technological packages, such as new seeds, soil correction techniques, and improved production practices, but later expanded to also develop high-yielding and disease-resistant crops. EMBRAPA was an institutional innovation designed to respond to a diversity of agricultural needs over a vast geographical area. It has a number of distinctive features that include the use of a public corporation model; national scale of operations in nearly all states; geographical decentralization; specialized research facilities (with 38 research centers, 3 service centers, and 13 central divisions); emphasis on human resource development (74% of 2,200 researchers have doctoral degrees while 25% have master's training); improvements in remuneration for researchers; and a strategic outlook that emphasizes science and innovation as well as commercialization of research results.[28]

EMBRAPA, a federal institution, is complemented by state-based research agencies within the National System for Agriculture Research and Innovation (SNPA), which have also assisted with the promotion and development of agribusiness innovation in the last decades. Implementation of the SNPA has paved the way for "a strengthening of agricultural R&D capacity in Brazil, with improved infrastructure, human capacity, management mechanisms and support policies on a national scale."[38] These government research agencies are supported by a larger, complex agricultural research system that encompasses public institutes, universities, public companies and NGOs, which helps foster knowledge exchange and drives innovation. Because of this consolidation, it is estimated that in 2006, Brazil was responsible for 41% of the $3.0 billion invested in agricultural research by the 27 countries in Latin America and the Caribbean.[39] The success of the Brazilian agriculture story should motivate African countries with agricultural growth prospects to adopt similar institutional reforms to support innovation, growth, and development in the sector.

Conclusion

Agricultural innovation has the potential to transform African agriculture, but only if strong structures are put in place to help create and disseminate critical best practices and technological breakthroughs. In much of Africa, linkages between farmers, fishermen, and firms and universities, schools, and training centers could be much stronger. New telecommunications technologies such as mobile phones have the potential to strengthen linkages, but cluster theory suggests that geography will continue to matter, regardless of new forms of communication. Groups that are closer physically, culturally, and socially are more likely to trust one another, exchange information and assets, and enter into complex cooperative production, processing, financing, marketing, and export arrangements.

Local, national, and regional authorities must carefully assess where clusters may prove most successful and must lay out clear plans for cluster development, which can take years if not decades. Local authorities should focus on identifying potential areas and industries for successful clusters. National governments should focus on providing the knowledge, personnel, capital, and regulatory support necessary for cluster formation and growth. And regional authorities should focus on linking national clusters to one another and to key related global institutions. Throughout these processes, public and private institutions must work cooperatively, with the latter being willing to transfer knowledge, funding, and even personnel to the private sector in the early stages of cluster development.

To promote innovation, the public sector could further support interactions, collective action, and broader public-private partnership programs. The country studies suggest that from a public sector perspective, improvements in agricultural innovation system policy design, governance, implementation, and the enabling environment will be most effective when combined with activities to strengthen innovation capacity. Success stories in which synergies were created by combining market-based and knowledge-based interactions and strong links within and beyond the value chain point to an innovation strategy that is holistic in nature and that focuses, in particular, on strengthening the interactions between key public, private, and civil society actors.

5

ENABLING INFRASTRUCTURE

Enabling infrastructure (public utilities, public works, transportation, and research facilities) is essential for agricultural development. Infrastructure is defined here as facilities, structures, associated equipment, services, and institutional arrangements that facilitate the flow of agricultural goods, services, and ideas. Infrastructure represents a foundational base for applying technical knowledge in sustainable development and relies heavily on civil engineering. This chapter outlines the importance of providing an enabling infrastructure for agricultural development.[1] Modern infrastructure facilities will need to reflect the growing concern over climate change. In this respect, the chapter will focus on ways to design "smart infrastructure" that takes advantage of advances in the engineering sciences, as well as ecologically sound systems design. Unlike other regions of the world, Africa's poor infrastructure represents a unique opportunity to adopt new approaches in the design and implementation of infrastructure facilities.

Infrastructure and Development

Poor infrastructure and inadequate infrastructure services are among the major factors that hinder Africa's sustainable development. This view has led to new infrastructure development approaches.[2] Without adequate infrastructure, African countries will not be able to harness the power of science and innovation to meet sustainable development objectives and to be competitive in international markets. Roads, for example, are critical for supporting rural development. Emerging evidence suggests that in some cases low-quality roads have a more significant impact on economic development than high-quality roads. In addition, all significant scientific and technical efforts require reliable electric power and efficient logistical networks. In the manufacturing and retail sectors, efficient transportation and logistical networks allow firms to adopt process and organizational innovations, such as the just-in-time approach to supply chain management.

Infrastructure promotes agricultural trade and helps integrate economies into world markets. It is also fundamental to human development, including the delivery of health and education services. Infrastructure investments further represent untapped potential for the creation of productive employment. For example, it has been suggested that increasing the stock of infrastructure by 1% in an emerging country context could add 1% to the level of GDP. But in some cases the impact has been far greater: the Mozal aluminum smelter investment in Mozambique not only doubled the country's exports and added 7% to its GDP, but it also created new jobs and skills in local firms.

Reducing public investment in infrastructure has been shown to affect agricultural productivity. In the Philippines, for example, reduction in investment in rural infrastructure led to reductions in agricultural productivity.[3] This decline in investment was caused by cutbacks in agricultural investments writ large, as well as by a shift in focus from rural infrastructure and agricultural research to agrarian reform, environment, and natural resource management. Growth in Philippine agriculture in the 1970s was linked to increased investments in infrastructure, just as declines in the same sector in the 1980s were linked to reduced infrastructure investment (caused by a sustained debt crisis).

Evidence from Uganda suggests that public investment in infrastructure-related projects has contributed significantly to rural development.[4] Uganda's main exports are coffee and cotton; hence, the country depends heavily on its agricultural economy. Political and economic turmoil in the 1970s and 1980s in Uganda led to the collapse of the economy and agricultural output. Reforms in the late 1980s allowed Uganda to improve its economic growth and income distribution. In spite of economic growth ranging between 5% and 7%, the growth of the agricultural sector has been very low, averaging 1.35% per annum. Even if the Ugandan government has made great strides in welfare improvement, the rural areas still remain relatively poor. In addition, due to the disparity between male and female wages in agriculture, women are more affected by poverty than men.

The Ugandan government has been spending on a wide variety of sectors, including agriculture, research and development, roads, education, and health (data in other sectors such as irrigation, telecommunications, and electricity are limited). Previous studies have mostly measured the effectiveness of government spending based on budget implementation.

Government spending on agricultural research and extension improved agricultural production substantially in Uganda. Growth in agricultural labor productivity, rural wages, and nonfarm employment have emerged as important factors in determining rural poverty, so much so that the public expenditure on agriculture outweighs the education and health effect. Investment in agriculture has been shown to increase food production and to reduce poverty. Roads linking rural areas to markets also serve to improve agricultural productivity and increase nonfarm employment opportunities and rural wages. Having a high HIV/AIDS prevalence, a large share of Uganda's health expenditure goes toward prevention and treatment. Despite the high expenditure in health services, there does not seem to be a high correlation between health expenditure and welfare improvement.

Infrastructure and Agricultural Development

Transportation

Reliable transportation is absolutely critical for growth and innovation in African agriculture and agribusiness. Sufficient roads, rail, seaports, and airports are essential for regional trade, international exports, and the cross-border investments that make both possible. Innovation in other areas of agriculture, such as improved genetic material, better access to capital, and best farming practices, will produce results only if farmers and companies have a way to get their products to market and get critical inputs to farms.

Transportation is a key link for food security and agribusiness-based economic growth. Roads are the most obvious and critical element, but modern seaports, airports, and rail networks are also important, particularly for export-led agricultural innovation, such as cut flowers and green beans in Kenya, neither of which would be possible without an international airport in Nairobi. To that end, many African countries have reprioritized infrastructure as a key element in their agricultural development strategies. This section will examine the role that roads have played in China's rural development and poverty alleviation, as well as two cases in African transportation investment: Ghana's Rural Roads Project and Mali's Bamako-Sénou Airport Improvement Project.

Ghana's Rural Roads Project "is expected to open new economic opportunities for rural households by lowering transportation costs (including travel times) for both individuals and cargo to markets and social service delivery points." The project will include new construction, as well as the "improvement of over 950 kilometers of feeder roads, which, along with the trunk roads, will benefit a total population of more than 120,000 farming households with over 600,000 members. These activities will increase annual farm incomes from cultivation by US$450 to about US$1,000. For many of the poor, the program will represent an increase of one dollar or more in average income per person per day." In addition to sparking growth in agriculture, the feeder roads will also help "facilitate transportation linkages from rural areas to social service networks (including, for instance, hospitals, clinics, and schools)."[5]

Landlocked countries such as Rwanda suffer economically from poor transport infrastructure, spending as much as 84% more than coastal countries to export commodities to international markets. A $13.5 billion project is underway in East Africa to develop a 1,824-mile rail infrastructure line connecting Kigali to the port of Mombasa. The railway that currently exists on this path dates from the colonial era, and has gauges too narrow for modern freight trains. The Mombasa-Kigali rail link will connect Mombasa, Nairobi, and Kisumu in Kenya, to Kampala, Uganda, before connecting to Kigali and four other more rural Rwandan towns. The line will be primarily designed for cargo traveling at 50 miles per hour, but can also accommodate passengers. Agricultural exports from Rwanda, like coffee and tea, will feature heavily in the freight composition; on the trip from Mombasa to Kigali, the railway is expected to carry machinery and other manufactured goods. The introduction of an alternative pathway for transporting cargo to and from ports is not only expected to reduce costs for business, but also to improve the longevity of roads by reducing the burden of heavy trucks.[6]

The Airport Improvement Project will expand Mali's access to markets and trade through improvements in the transportation infrastructure at the airport, as well as better management of the national air transport system. However, Mali is landlocked and heavily dependent on inadequate rail and road networks and port facilities in countries whose recent instability has cost Mali dearly. Before the outbreak of the Ivorian crisis, 70% of Malian exports were leaving via the port of Abidjan. In 2003, this amount dwindled to less than 18%. Mali cannot control overland routes to international and regional markets. Therefore, air traffic became Mali's lifeline for trans-portation of both passengers and export products.

Malian exports are predominantly agriculture based and depend on rural small-scale producers, who will benefit from increased exports in high-value products such as mangoes, green beans, and gum arabic. The Airport Improvement Project is intended to remove constraints to air traffic growth and increase the airport's efficiency in both passenger and freight handling through airside and landside infrastructure improvements, as well as the establishment of appropriate institutional mechanisms to ensure effective management, security, operation, and maintenance of the airport facilities over the long term.[7]

In response to requirements for safety and security audits by the International Civil Aviation Organization and the US Federal Aviation Administration, Mali was in the process of restructuring and consolidating its civil aviation institutional framework. One major result has been the establishment of the new civil aviation regulatory and oversight agency in December 2005, which gained financial and administrative independence. The Airport Improvement Project will reinforce the agency by providing technical assistance to establish a new organizational structure, administrative and financial procedures, staffing and training, and provision of equipment and facilities. Additionally, the project will rationalize and reinforce the airport's management and operations agency by providing technical assistance to establish a model for the management of the airport and the long-term future status and organization of agency.

Since 1985 China has given high priority to road development, particularly high-quality roads such as freeways. While the construction of high-quality roads has taken place at a remarkably rapid pace, the construction of lower-quality and mostly rural roads has been slower. Benefit-cost ratios for lower-quality roads (mostly rural) are about four times *larger* than those for high-quality roads when the benefits are measured in terms of national GDP.[8]

In terms of welfare improvement, for every yuan invested, lower-quality roads raised far more rural and urban poor people above the poverty line than did high-quality roads. Without these essential public goods, efficient markets, adequate health care, a diversified rural economy, and sustainable economic growth will remain elusive. Effective development strategies require good infrastructure as their backbone. The enormous benefit of rural roads in China likely holds true for other countries as well. Investment in rural roads should be a top priority for reducing poverty, maximizing the positive effects of other pro-poor investments, and fostering broadly distributed economic growth. Although highways remain critical, lower-cost, often lower-quality rural feeder roads are of equal and in some cases even greater importance.

"As far as agricultural GDP is concerned, in today's China additional investment in high quality roads no longer has a statistically significant impact while low-quality roads are not only significant but also generate 1.57 yuan of agricultural GDP for every yuan invested. Investment in low-quality roads also generates high returns in rural nonfarm GDP. Every yuan invested in low-quality roads yields more than 5 yuan of rural nonfarm GDP."[9] Low-quality roads raise more poor people out of poverty per yuan invested than high-quality roads, making them a win-win strategy for growth in agriculture and poverty alleviation. In Africa, governments can learn from the Chinese experience and ensure that their road programs give adequate priority to lower-quality and rural feeder roads.

Energy

To enhance agricultural development and to make progress in value-added agro-processing, Africa needs better and more consistent sources of energy. Rolling blackouts are routine in much of western, central, and eastern Africa, and much of Africa's power generation and transmission infrastructure needs repair or replacement. What Africa lacks in adequate deployment, however, it makes up for in potential. Africa is endowed with hydro, oil, natural gas, solar, geothermal, coal, and other resources vast enough to meet all its energy needs. Nuclear energy is also an option. The hydro potential of the Democratic Republic of the Congo is itself enough to provide three times as much power as Africa currently consumes.

The first step to improved power generation and transmission is to repair and upgrade Africa's existing energy infrastructure. Many African countries are operating at less than half their installed potential due to inadequate maintenance and operation. Connecting rural areas to national grids can in some cases be cost prohibitive, so governments must also look for innovative solutions such as wind, solar, biomass, and geothermal to provide power at the small farm level. Finally, while countries will undoubtedly look first within their own borders for resources, advanced energy planning should also consider that the most affordable and reliable power may be in neighboring states. Large power generation schemes may also require cooperative agreement on resource management and funding from a host of African and international sources. Cross-border energy networks could help create a common market for energy, spur investment and competition, and lead to a more efficient path of enhanced energy infrastructure.

An example of such a regional energy system is the West African Power Pool (WAPP). Under an agreement signed by 14 ECOWAS members in 2000, countries plan to "develop energy production facilities and interconnect their respective electricity grids. . . . ECOWAS estimates that 5,600 kilometers of electricity lines connecting segments of national grids will be put in place. About US$11.8 billion will be needed for the necessary power lines and new generating plants. This infrastructure would give the ECOWAS subregion an installed capacity of 10,000 megawatts" and, critically for agro-processing and business investment, dramatically increase not just the amount but also the reliability of electricity in West Africa.[10]

The West African Power Pool aims to establish a power-pooling mechanism between the national power companies of the ECOWAS member states; doing so requires a unified legal and regulatory framework. Crucial policy innovations in this framework include the risk-free exchange of electricity between countries in the context of a transparent pricing agreement, and bind member states to help one another in the case of power system calamity to avert collapse. The West African Power Pool focuses on giving member states access to reliable and low-cost energy supplies from hydropowered and gas-fired plants.

The West African Power Pool organization has been created to integrate the national power system operations into a unified regional electricity market—with the expectation that such a mechanism would, over the medium to long term, assure the citizens of ECOWAS member states a stable and reliable electricity supply at affordable costs.[11] This will create a level playing field, facilitating the balanced development of the diverse energy resources of ECOWAS member states for their collective economic benefit, through long-term energy sector cooperation, unimpeded energy transit, and increased cross-border electricity trade. The major sources of electricity under the power pool would be hydroelectricity and gas to fuel thermal stations. Hydropower would be mainly generated on the Niger (Nigeria), Volta (Ghana), Bafing (Mali), and Bandama (Côte d'Ivoire) rivers. The World Bank has committed a $350-million line of credit for the development of the WAPP, but a billion more is needed in public and private financing.

Most of the power supply in Africa is provided by the public sector. There is growing interest in understanding the ability of independent power projects (IPPs) in Africa by evaluating a project's ability to produce reliable and affordable power as well as reasonable returns on investment.[12] In the context of their individual markets, the 40 IPPs under consideration have played a complementary role to state-owned power projects, filling gaps in supply. It was also hoped that, once established, these private entities would introduce competition into the market.

Evidence suggests that there is a dichotomy between relatively successful IPPs, situated mainly in the northern African nations of Egypt, Morocco, and Tunisia, and the sub-Saharan examples, in Ghana, Kenya, Nigeria, and Tanzania, which have been less successful. A wide variety of country-level factors, including investment climate, policy frameworks, power sector planning, bidding processes, and fuel prices, all impacted outcomes for these various IPPs. Despite their private nature, ultimately it is the perceived balance of commitment between sponsors and host-country governments that plays one of the largest roles in the outcome of the IPP. A leading indicator of imbalance is frequent and substantial contract changes.

The presence of a favorable climate for investment influenced the outcome of the IPPs. In the more successful North African examples, Tunisia carried an investment grade rating, while Egypt and Morocco were both only one grade below investment grade. In contrast, of those nations located in sub-Saharan Africa, none received an investment grade rating. The great demand for IPPs in Africa at the time meant that those with superior investment profiles were able to attract more investors and had a basis for negotiating a more balanced contract.

Few of the nations in question have established a clear and coherent policy framework within which an IPP could sustainably operate. The soundest policy frameworks again are found in the north, with Egypt, which contains 15 IPPs, being the strongest. This policy framework features a clearly defined government agency in the Egyptian Electricity Authority, which has authority over the procurement of IPPs, the allocation of new generation capacity, and the ability to set benchmarks to increase competition among public facilities. Kenya set itself apart in this context with the establishment of an independent regulator—the Electricity Regulatory Board—which has helped to significantly reduce power purchase agreement charges, to set tariffs, and to mediate the working relationship between the public and private sectors. Evidence suggests that if a regulator is established prior to negotiation of the IPP, and acts in a transparent, fair, and accountable manner, this office can have a significantly positive effect on the outcomes for the host country and investor.

A coherent power sector plan follows from a strong policy framework and includes setting a reliability standard for energy security, supply and demand forecasts, a least-cost plan, and agreements on how new generation will be divided between public and private sectors. It is equally important that these functions are vested in one empowered agency. Failure to meet these goals is apparent in the examples of Tanzania (Songo Songo), Kenya (Westmont plant, Iberafrica plant), Nigeria (AES Barge), and Ghana, which fast-tracked IPPs to meet intermediate power shortages in the midst of drought conditions. The results were unnecessary costs and time delays for all, and in the case of the Nigerian and Ghanaian facilities, an inability to efficiently establish power purchase agreements.

The main lesson learned here is that without a strong legislative foundation and coherent planning, contracts were unlikely to remain intact. Instability of contracts was widespread across the cases studied, and though they did not necessarily deal a death blow to the project, renegotiations always came at a further cost.

US President Barack Obama's signature policy achievement in Africa is likely to be the launch of the $7 billion Power Africa initiative. The project, which was announced during Obama's trip to the continent in 2013, aims to capitalize on and coordinate action between US development agencies, African governments, and private investors to double access to electricity by 2018. Power Africa focuses on Ethiopia, Ghana, Kenya, Liberia, Nigeria, and Tanzania. The US government is providing

$7 billion in financial support and guarantees; more signifi-
cantly, US investors have committed $14 billion. These various
parties work together on the specifics of given power deals.
For example, the US government helped convince Tanzania
to extend its standard power purchasing agreement from 15
to 25 years to help obtain private-sector financing. While the
magnitude of the energy deficit on the continent dwarfs Power
Africa, the US government aims to promote the initiative as a
new way forward in foreign aid, bringing together public and
private actors for maximum effect.[13]

Irrigation

Investment in water management is a crucial element of successful agricultural development and can be broken into two
principal areas: policy and institutional reforms, on the one
hand, and investment, technology, and management practices,
on the other. Water is also a critical input beyond agriculture,
and successful irrigation policies and programs must take
into account the key role of water in energy production, public
health, and transportation. For small farmers, low-cost technology is available, and there are cost-efficient technical solutions, even in some of Africa's most difficult and arid regions.
Despite the availability of these technologies, Africa has not
seen widespread adoption of these techniques and technologies. Part of the problem is the availability of finance and the
slow spread of knowledge, but equally important is the role of
government regulations and subsidies.

Successful strategies for improved water management and
irrigation must therefore not only focus on new technologies
but also on creating policies and regulations that encourage
investment in irrigation, not just at the farm but also at the
regional level. Access to reliable water supplies has proven a
key determinant, not just in the enhancement of food security,
but also in farmers' ability to climb higher up the value chain
toward cash crops and processed foods. Innovative farmers

involved in profitable agro-export may represent a new constituency for the stewardship of water resources, as they earn significantly higher incomes per unit of water than conventional irrigators. The analysis will focus first on innovation in water management practices, technology, and infrastructure (including examples from Mali, Egypt, and India). The final section will also address key water policy and institutional reforms necessary to create an environment in which governments, international institutions, NGOs, and private businesses will be encouraged to make investments in irrigation infrastructure.

Begun in 2007, the Alatona Irrigation Project will provide a catalyst for the transformation and commercialization of family farms, supporting Mali's national development strategy objectives to increase the contribution of the rural sector to economic growth and to help achieve national food security. Specifically, it will increase production and productivity, improve land tenure security, modernize irrigated production systems, and mitigate the uncertainty from subsistence rain-fed agriculture, thereby increasing farmers' incomes. The Alatona Irrigation Project will introduce innovative agricultural, land tenure, credit, and water management practices, as well as policy and organizational reforms aimed at realizing the Office du Niger's potential to serve as an engine of rural growth for Mali. This project seeks to develop 16,000 hectares of newly irrigated lands in the Alatona production zone of the Office du Niger, representing an almost 20% increase of "drought-proof" cropland.

A project in Benin conducted by Stanford University and the Solar Electric Light Fund found that solar-powered drip irrigation systems can improve rural incomes and downstream development indicators like nutritional intake. A small pilot project found that one solar-powered irrigation system supported on average 1.9 metric tonnes of produce per month. This system enables farmers to move beyond a season-limited growing season, and to begin producing a diversified selection of value-added crops. Villagers benefiting from the solar irrigation system reported a 17% decrease in food security. Challenges to the widespread implementation of such promising systems include not only the relatively high up-front costs of $18,000 per system, but also the nearly $6,000 annual maintenance costs.[14]

About 85% of the Nile River's water is used for irrigation. Egypt depends almost entirely on water from the Nile, drawing on 95% of the available resources for the country.[15] Egypt's water ministry has responded to the mismatch between water supply and demand by improving water management throughout the water management system and on-farm applications by strategically sizing water infrastructure to optimize the use of capital outlays, applying technical innovations to save water and money. The project aims to improve farmers' annual income by about 15% annually while also seeing a water savings of 22%.[9]

Sugarcane cultivation requires significant water resources, but in much of India it has been cultivated using surface irrigation, where water use efficiency is very low (35%–40%), owing to substantial evaporation and distribution losses.[16] A recent study of sugarcane cultivation in Tamil Nadu, India, has shown that using drip irrigation techniques can increase productivity by approximately 54% (30 tons per acre) and can cut water use by approximately 58% over flood irrigation. Unlike surface methods of irrigation, under drip methods, water is supplied directly to the root zone of the crops through a network of pipes, a system that saves enormous amounts of water by reducing evaporation and distribution losses. Since water is supplied only at the root of the crops, weed problems are less severe and thus the cost required for weeding operations is reduced significantly. The system also requires little, if any, electricity.

Although new and larger studies are necessary, initial analysis suggests that investment in drip irrigation in Indian sugarcane cultivation is economically viable, even without subsidy, and may also be applicable in Africa, where many farmers have limited or no access to electrically powered irrigation, water resources are increasingly threatened by climate change and environmental degradation, and less than 4% of the arable land is currently irrigated.[17]

Further, the present net worth indicates that in many cases farmers can recover their entire capital cost of drip irrigation from first-year income without subsidy. Despite these gains, two impediments must be overcome for drip irrigation to be more widely used, not just in India, but in much of the developing world. First, too few farmers are aware of the availability and benefits of drip irrigation systems, which should be demonstrated clearly and effectively through a quality extension network. Second, despite the quick returns realized by many farmers using drip irrigation, the systems require significant capital up front. Banks, microcredit institutions, companies, and governments will need to consider providing credit or subsidies for the purchase of drip irrigation.

The total cultivated land area of the Common Market for Eastern and Southern Africa amounts to some 71.36 million hectares. Of this, only about 6.48 million hectares are irrigated, representing some 9% of the total cultivated land area. Besides available land area for irrigation, the region possesses enormous water resources and reservoir development potential to allow for expansion. Of the world's total of 467 million hectares of annualized irrigated land areas, Asia accounts for 79% (370 million hectares), followed by Europe (7%) and North America (7%). Three continents—South America (4%), Africa (2%), and Australia (1%)—have a very low proportion of global irrigation. COMESA could contribute significantly to agricultural food production and poverty alleviation through expanding the land under irrigation and water management under rain-fed farming to effect year-round crop and livestock production.

COMESA has recently made assessments through Comprehensive Africa Agriculture Development Programme reports involving some representative countries with respect to agriculture production options, and concluded that regional economic growth and food security could be accelerated through investment in irrigation and agriculture water management. Agriculture water-managed rain-fed yields are similar to irrigated yields and always are higher than rain-fed agriculture yields. This scenario builds a watertight case for promoting or expanding irrigated land in COMESA.

The best solution to poverty and hunger alleviation is to provide people with the means to earn income from the available resources they have. Small-scale irrigation development, coupled with access to long-term financing, access to markets, and commercial farming expertise by producers, will go a long way in achieving food security and overall economic development. COMESA has created an agency called the Alliance for Commodity Trade in Eastern and Southern Africa to implement practical investment actions by engaging public-private sector partnerships. In the areas of irrigation and agriculture water management, COMESA has begun to implement a number of important activities.

Accelerated adoption of appropriate small-scale irrigation technologies and improved use and management of agriculture water will facilitate increased agricultural production and family incomes. The rain-fed land area will require agriculture water management strategies such as conservation agriculture, which enhances production. Appropriate investment in field systems for irrigation with modest investments will help smallholder farmers adopt irrigation technology, whereas the majority who practice rain-fed agriculture would improve agriculture productivity by managing rainwater through systems such as conservation agriculture technology. COMESA is embarking on reviewing the policy and legal framework in water resources management programs including trans-boundary shared water resources management policies under CAADP. This will include actions toward adaptation by member states of regional water resources management policies.

COMESA is working with regional and international organizations such as Improved Management of Agriculture Water in Eastern and Southern Africa, East African Community, Southern African Development Coordination, Intergovernmental Authority for Development, International Water Management Institute, and Wetland Action-UK in creating awareness in regional sustainable water resources management by creating and strengthening water dialogue platform and communication strategies. Through the Alliance for Commodity Trade in Eastern and Southern Africa, COMESA will help develop regional water management information systems observation networks so as to enhance mapping for water-harvesting resources and water utilization in COMESA.

To realize the benefits of irrigation and agriculture water management, COMESA is promoting investment in the following areas: reservoir construction for storage of water to command an expansion of land area under irrigation by 30% in five years; inland water resources management of watershed basins in the COMESA region, including policy and legal frameworks in trans-boundary shared water resources management, harmonizing shared water resources policies to optimize utilization, strengthening regional institutions involved in water resources management, and establishment of a regional water resources management information system; building capacity and awareness for sustainable water resources utilization and management for agricultural food production; rapid expansion of terraces for hilly irrigation in some member states; and promotion and dissemination of appropriate irrigation and agriculture water management technology transfer and adoption. These include smallholder irrigation infrastructure.

Telecommunications

Access to timely weather, market, and farming best practices information is no less important for agricultural development than access to transport infrastructure, regular and efficient irrigation, and energy. In Africa, as in much of the developing world, innovations in telecoms offer the potential to bring real change directly to the farm level long before more timely and costly investment in fixed infrastructure. Mobile phone penetration rates now exceed those of landlines, and the industry is growing at an average annual rate of over 50% in the region. Mobile phone ownership in Africa increased from 54 million in 2003 to nearly 500 million in 2010. The penetration rate is now over 50%. Ownership rates underrepresent actual usage, however, as many small vendors offer mobile access for calls or text messages. Even in rural areas, mobile penetration rates have now reached close to 42%. Mobile phones are becoming increasingly important tools in agricultural innovation, where they have been used to transfer and store money, check market prices and weather information, and even share farming best practices.

The case of India's e-Choupal (*choupal* is Hindi for a gathering place) illustrates the increasingly important role that telecommunications can play in African agricultural innovation.[18] ITC, an Indian private company with annual turnover of US$7 billion, brings Internet access and computers to rural villages through its e-Choupal initiative. It places computers in rural villages for a setup cost of $3,000–$6,000, and annual maintenance of $100. Most often the computer is inside one farmer's home. The computer host is responsible for some operating costs, and is bound by oath to serve the entire community; that farmer is compensated through commissions on transactions. The computer is hooked up to the Internet through a satellite connection or phone lines.

The computers have come to serve as community centers and hubs for the exchange of information. A single computer serves an average of 600 farmers, with a network reaching into neighboring villages up to 5 kilometers away. By 2010, there were 6,500 e-Choupals serving over 4 million farmers. The farmers use the system for free. The e-Choupals bring information about price trends for agricultural commodities, as well as closing prices in nearby markets, and new farming techniques. The farmers can order fertilizer, seeds, or other agricultural goods through ITC or its business affiliates at lower prices than farmers will find from village traders. When farmers harvest their crops, ITC also offers to buy the produce at the previous day's closing rates, transports it to a processing center, grades it for quality, and offers bonuses for high-quality crops.

Not only is the corporate effort yielding positive development outcomes, bringing information and transparency to rural farmers, but it also has become highly profitable for ITC as an e-commerce platform designed specifically for rural populations. The farmers are gaining from the integrated, transparent, and faster sale system through ITC, as well as lower prices for inputs, and access to information on farming techniques. ITC overall has net procurement costs 2.5% lower than it would without the e-Choupal system. ITC recovers equipment costs from e-Choupal within a year of operation, and rates the overall system as profitable.[19]

The availability of weather information systems for farmers is also emerging as a critical resource. Although advances in irrigation infrastructure and technology are lowering farmers' dependency on weather, a second avenue to advance agricultural development is through more accurate and accessible weather information. To address the gap in accurate, timely, and accessible weather information in Africa, the Global Humanitarian Forum, Ericsson, World Meteorological Organization, Zain, and other mobile operators have developed a public-private partnership to (1) deploy up to 5,000 automatic weather stations in mobile network sites across Africa, and (2) increase dissemination of weather information via mobile phones to users and communities—including remote farmers and fishermen.

Zain will host the weather equipment at mobile network sites being rolled out across Africa, as achieving the target

of 5,000 sites will require additional operator commitment
and external financing. Mobile networks provide the necessary connectivity, power, and security to sustain the weather
equipment. "Through its Mobile Innovation Center in Africa,
Ericsson will also develop mobile applications to help com-
municate weather information developed by national meteorological and hydrological services. . . . Mobile operators
will maintain the automatic weather stations and assist in the
transmission of the data to national meteorological services.
The initial deployment, already begun in Zain networks, focuses on the area around Lake Victoria in Kenya, Tanzania,
and Uganda. The first 19 stations installed will double the
weather monitoring capacity of the lake region."[20]

Infrastructure and Innovation

One of the most neglected aspects of infrastructure investments is their role in stimulating technological innovation. Development of infrastructure in a country is often not enough
to create sustained economic growth and lifestyle convergence
toward that of developed countries. Technological learning is
very important to a country's capacity to maintain current infrastructure and to become competitive. In the first model of
technology transfer, state-owned or privatized utility firms
couple investment in public infrastructure with technological
training programs, usually incorporated into a joint contract
with international engineering firms. This type of capacity
building lends itself to greater local participation in future infrastructure projects, both within and outside the country.

The effectiveness of a comprehensive collaboration with
foreign companies to facilitate both infrastructure building
and technology transfer is seen in South Korea's contract
with the Franco-British Consortium Alstom. The Korean
government hoped to develop a high-speed train network
to link Seoul with Pusan and Mokpo. The importance
of the infrastructure itself was undeniable—the Korean

Train Express (KTX) was meant to cross the country, going through a swath of land responsible for two-thirds of Korea's economic activity. In anticipation of the project, officials projected that by 2011, 120 million passengers would be using the KTX per year, leading to more balanced land development across South Korea.

However, while the project had the potential to increase economic activity and benefit the national industries in general, the project's benefits lay in the opportunity "to train its workforce, penetrate a new industrial sector, and potentially take the lead in the high-speed train market in Asia."[21] In other words, Korea sought to obtain new technologies and the capacity to maintain and operate them. Under the contract with Alstom, which was finalized after 20 years of discussion, Alstom provided both the high-speed trains and railways that would help connect Seoul and Pusan and the training to help South Korea build and maintain its own trains.

From the beginning of negotiations, technology transfer was an important factor for the South Korean government. In 1992, bidding between Alstom, Siemens (a German group), and Mitsubishi (a Japanese group) commenced. After the bids were significantly slashed, Korean officials let it be known that in addition to price, financial structures and technology transfers would be major criteria during the final selection. It was in this category that Alstom successfully outbid the other consortia and won the contract, which specified that half of all production would occur in Korea, with 34 of the trains to be built by Korean firms. This would give Korea both production revenue and the experience of building high-speed trains—with the goal of one day exporting them. The contract also stipulated that 100% of Alstom's TGV (Train à Grande Vitesse) technology would be transferred to the 15 Korean companies that were to be involved in the project. Such technologies include industrial planning, design and development of production facilities, welding, manufacturing, assembly and testing carried out through operating and maintenance training, access to important documents and manuals for technical assistance, and maintenance supervision.

While the overall benefits of technology transfer are clear, more technologically advanced countries face some risks. One risk, known as the boomerang effect, affects the company that is transferring the technology—Alstom in this case. By giving the technological knowledge to South Korean companies, Alstom runs the risk of essentially creating its own competitor. This risk is especially high in this case because Alstom has transferred 100% of its TGV technology and 50% of the production to Korea. Low labor costs, weak contractual constraints, and Korea's known tendency to disregard intellectual property rights increase this risk. Other risks include unexpected shifts in economic conditions, currency devaluations, questionable competitive practices, hurried local production, lengthy and cumbersome administrative procedures, restrictive foreign payment rules, management weaknesses, and frail partnership involvement.

While these risks are indeed significant, they should not deter such agreements between countries. There are many ways to decrease such risks. For example, to make sure that payments are timely and that intellectual property rights are upheld, the company of interest should create a detailed contract with large penalties and disincentives for any violations.

Another step that should be taken is to maintain strong research and development projects to ensure that one's technology will always be superior. A good way to avoid the boomerang effect is to establish long-term relationships, such as Alstom established with South Korea. A similar method of preventing the boomerang effect is to establish partnerships with local manufacturers. Finally, Alstom took strides to collaborate with established competitors, like the formation of EUROTRAIN with Siemens, to increase penetration into new markets.

And despite the numerous risks, training and technology transfer did not result in a loss for Alstom, for benefits included numerous cash payments, dividends, and income from giving access to its technology, selling equipment parts, and establishing separate ventures with South Korean companies. Additionally, the project gave the company the opportunity to show the exportability of TGV to Asian markets. In particular, the reliability of Alstom's products and procedures was demonstrated in the partnership, making other countries more likely to work with the company. Experience in the Korean market, competitive advantages with respect to European countries, and new business opportunities were other advantages that increased Alstom's market share in Asia. Increased flexibility and experience with international markets, as well as decentralized management, also benefited Alstom. Finally, Alstom's technology became the international standard, leading to enormous competitive advantages for that company.

To facilitate the transfer, development, and construction of the high-speed railway system, the South Korean government created the Korean High-Speed Rail Construction Authority (KHRC), whose mandate was to construct such systems at home and abroad, to research and find ways to improve the technology, and to oversee commercialization along the railway line. Issues with the project were soon revealed; after two tragedies—the collapse of the Songsu Bridge and of a large store in Seoul—Korean officials began to doubt its civil engineering capabilities. For this reason, KHRC decided to hire foreign engineers. After project delays and other issues, the last section of railway track from Taegu to Pusan was canceled and the building of 34 trains was postponed. However, after renegotiations, construction recommenced.

Other issues show the difficulty of such a project collaboration. A rift developed between France and South Korea due to a withdrawn agreement between two companies. Further, the TGV was unable to function in Korea during an unusually cold winter in 1996–1997, drawing questions and critiques from the Korean press. An economic crisis, which caused an abrupt depreciation of the Korean won against the US dollar, made the purchase of goods and services from foreigners more expensive. A final rift was created by the election of President Kim Dae Jung in 1997, who was a vocal opponent of the KTX project. As seen here, exogenous interactions between the countries of interest can greatly affect the attempts to collaborate in technological transfer.

Despite the many risks of international technology transfer, the benefits far outweigh the costs. For example, by 2004, 100% of the TGV technology was transferred to South Korea. Despite initial setbacks, ridership has increased greatly at the expense of other modes of transportation. More lines are expected to be built, and the success of the technology transfer has become apparent through the construction of the HSR-350x Korean-made train and the order of 19 KTX-II train sets in 2006 from Hyundai Rotem. It is claimed that these trains use 87% Korean technology. As for Alstom, their success is evident in the numerous contracts they have negotiated in the Asian markets.

In conclusion, the KTX project demonstrates how technology transfer can help developing countries to obtain advanced capabilities to build and develop infrastructure, leading to increased economic growth and productivity. The South Korean example serves as a model for African countries and applies to urban and rural projects alike. The lessons are particularly important considering the growing interest among African countries in investing in infrastructure projects.[22] While African countries will face their own unique issues, the KTX project illustrates costs and benefits that should be weighed in making such decisions and provides hope for new methods of technological dissemination. The tendency, however, is to view infrastructure projects largely in terms of their returns on investment and overall cost structure.[23] Their role in technological capacity building is rarely considered.[24] The growing propensity to want to leave infrastructure investments to the private sector may perpetuate the exclusion of public interest activities such as technological learning.[25]

One of the key aspects of the project was a decision by the government to set up the Korea Railroad Research Institute (KRRI). Founded in 1996, KRRI is the nation's principal railway research body. Its focus is improving the overall national railway system to maintain global competitiveness, with the goal of putting South Korea among the top five leaders in railway technology. It works by bringing together experts from academia, industry, and government.

Regional Considerations

Roads, water facilities, airports, seaports, railways, telecommunications networks, and energy systems represent just a portion of the web of national and regional infrastructure necessary for food security, agricultural innovation, and agriculture-based economic development. Countries and regions must create comprehensive infrastructure investment strategies that recognize how each area is linked to the next, and investments must in many cases pool regional resources and cross numerous international borders. Transportation infrastructure is critical to move inputs to farms and products to market; widespread and efficient irrigation is essential for increasing yields and crop quality; energy is a vital input, particularly for value-added food processing; and telecommunication is critical for the exchange of farming, market, and weather information. Alone, however, none of these investments will produce sustainable innovation or growth in agriculture. National and regional investment strategies will be needed to pool resources, share risks, and attract the private actors often critical to substantial investments in such ventures.

It seems obvious that roads would play a critical role in agricultural development, but they have often received inadequate investment. On-farm innovations are critical, but in many cases they depend on inputs that can only be delivered via roads, and they will be of very limited use if farmers have no way to reliably move their products to markets. Countries looking to improve their roads should carefully assess where their competitive advantages lie, identify which new or refurbished roads would best capitalize on those advantages, ensure that roads are placed within a broader plan for transportation infrastructure, and develop pre-construction plans for long-term maintenance.

Large roads and highways have garnered the bulk of capital and attention in much of the developing world, but smaller, lower-quality rural feeder roads often have significantly higher returns on investment—particularly in areas where major highways already exist. Learning from the Chinese experience, countries should carefully assess the relative return between larger highways and smaller rural-feeder roads, selecting the better investment.

National water policy and programs are notoriously Balkanized into fractious agencies and interest groups, often with competing objectives. This is a problem that countries across the world face, as is evidenced by the small American town of Charlottesville, Virginia. Charlottesville has no less than 13 separate water authorities representing its roughly 50,000 residents. As in the example of Egypt (a country with significant water resource pressures but a highly advanced water management system), an initial step to success is streamlining government regulation of water issues under a single national agency, or family of agencies. Water policy and programs should be coordinated at the national, not state, level, and must also look across borders to neighboring states as many key issues in water, including power generation, agricultural diversion, and water quality, are often closely linked to key issues up- or downstream.

Many African states already face water shortages, and the threat of global climate change may further stress those limited resources. Bringing new water assets online through large irrigation projects is important, but those resources are limited; more economical use of water is just as important, if not more so. Central to this goal are farming techniques that get "more dollar per drip." As we saw earlier with the case of India, drip irrigation can be one solution. To overcome the initial capital hurdle, governments, companies, and banks could consider subsidizing and/or providing loans for the purchase of initial equipment.

As with water, energy issues often transcend national borders. In many cases, the best location to produce or sell power may be outside a country's borders. Regional cooperation will be essential for unlocking much of Africa's energy-generation potential, as many projects will require far more investment than any one country can provide and involve assets that must span multiple national borders. To pool national resources and entice private capital for major energy products, regional organizations will need to help create strong, binding agreements to provide the necessary confidence not only to their member states but also to private companies and investors. The ECOWAS-led Western African Power Pool provides a good model for replication, but it is also an indicator of the high level of commitment and private capital that must be raised to push through large, regional power agreements.

Large power-generation and transmission schemes are critical to agricultural development but in some cases may prove too lengthy, costly, or difficult to have significant, timely impacts in remote rural areas. One way to complement these larger energy programs is to make additional investment in remote rural energy generation at the local or even farm level. Renewable technologies, including solar, wind, biogas, bioethanol, and geothermal, can be scaled for farms and small business and have the added advantage of requiring minimal transmission infrastructure and often a low carbon footprint. To encourage this production, governments could consider replicating Tanzania's Rural Energy Agency, which is funded by a small tax on sales from the national energy utility, as well as by partnerships with NGOs, foundations, foreign governments, and businesses.

The transfer of knowledge is nearly as important to agricultural innovation as the transfer of physical inputs and farm outputs. Telecoms can play a unique role in the transfer of farming best practices as well as critical market and weather information. Most of Africa's telecom infrastructure is owned by the private sector. As we have seen from cases in India, China, and Africa, private companies can play a key role in the development of telecoms as a tool for agricultural innovation. Governments and regional bodies should work with major telecom providers and agribusinesses to form innovative partnerships that provide profits to companies and concrete benefits such as enhanced farming knowledge transfer and market and weather information.

Mobile phone penetration rates are growing faster in Africa than anywhere in the world. Mobile phones and the cell tower networks on which they depend provide a unique platform for the collection and, even more important, the dissemination of key information, including farming best practices, market prices, and weather forecasts. To reach scale, Africa's regional organizations should engage their member states, key telecom businesses, and NGOs to harness existing technologies such as SMS (and next generation technologies such as picture messaging and custom applications for mobiles) to provide farmers with access to key agricultural, market, and weather information.

Conclusion

Infrastructure investment is a critical aspect of stimulating innovation in agriculture. It is also one of the areas that can benefit from regional coordination. Indeed, the various RECs in Africa are already increasing their efforts to rationalize and coordinate infrastructure investments. One of the lessons learned from other countries is the importance of linking infrastructure investment (especially in key areas such as transportation, energy, water, and telecommunications) to specific agricultural programs. It has been shown that low-quality roads connecting farming communities to markets could contribute significantly to rural development. An additional aspect of infrastructure investment is the need to use such facilities as foundations for technological innovation. One strategic way to achieve this goal is to link technical training institutions and universities to large-scale infrastructure projects. The theme of education, especially higher technical training, is the subject of the next chapter.

One of the most challenging aspects of investing in infrastructure is the high up-front costs. Making large investments ahead of demand creates a wide range of management and political risks. It is estimated that Africa will need nearly US$500 billion over the next decade to finance infrastructure projects. African countries will need to think creatively about how to secure the funding needed to support this level of investment. One option that remains unexplored is using the engineering and technical capacity in African militaries for infrastructure work in strategic areas such as transportation, irrigation, telecommunications, and energy. Countries such as Senegal have been mobilizing the military to contribute to infrastructure development and maintenance under clear management structures that do not open the door to political mischief. Uganda has been using the military for railway construction. These examples indicate the need to give careful consideration to the existing critical engineering capabilities of African militaries that can be harnessed to lay the foundations of economic security.

6

HUMAN CAPACITY

Education and human capacity building in Africa have many well-publicized problems, including low enrollment and completion rates. One of the most distressing facts about many African school systems is that they often focus little on teaching students to maximize the opportunities that are available to them in their own communities; rather, they tend to prioritize a set of skills that is less applicable to village life and that encourages children to aspire to join the waves of young people moving to urban areas. For some students, this leads to success, but for many more it leads to unfulfilled aspirations, dropout rates, and missed opportunities to learn crucial skills that will allow them to be more productive and have a better standard of living in their villages. It also results in nations passing over a chance to increase agricultural productivity, self-sufficiency, and human resources among their populations.

Education and Agriculture

African leaders have the unique opportunity to use the agricultural system as a driver for their economies and a source of pride and sustainability for their populations. About 36% of all African labor potential is used in subsistence agriculture. If that percentage of the population could have access to methods of improving their agricultural techniques, increasing production, and gaining the ability to transform agriculture into an income-earning endeavor, African nations would benefit in terms of GDP, standard of living, infrastructure, and economic stability. One way to accomplish this is to develop systems—both formal and informal—to improve farmers' skills and abilities to create livelihoods out of agriculture, rather than simply subsistence.

These systems start with formal schooling. Schools should include agriculture as a formal subject—from the earliest childhood experience to agricultural universities. They should consider agriculture an important area for investment and work to develop students' agricultural and technical knowledge at the primary and secondary levels. Universities should also consider agriculture an important research domain and should devote staff and resources to developing new agricultural techniques that make sense for their populations and ecosystems. University research needs to stay connected to the farmers and their lifestyles to productively foster agricultural growth.

Decisionmakers should also look for ways to foster human capacity to make agricultural innovations outside the traditional classroom. A variety of models incorporate this idea—from experiential and extension models to farmers' field schools, both discussed later in this chapter. Rural radio programs that reach out to farming communities and networks of farmers' associations spread new agricultural knowledge. In fact, there is a resurgence of radio as a powerful tool for communication.[1]

Governments and schools should treat agriculture as a skill to be learned, valued, and improved upon from early childhood through adult careers, instead of as a last resort for people who cannot find the resources to move to a city and get an industrial job. Valuing the agricultural system and lifestyle and trying to improve it take advantage of Africa's existing systems and capacities. In this way, many nations could provide significant benefits for their citizens, their economies, and their societies.

Nowhere is the missed opportunity to build human capacity more evident than in the case of women and agriculture in Africa. The majority of farmers in Africa are women. Women provide 70%–80% of the labor for food crops grown in Africa, an effort without which African citizens would not eat. Female farmers make up 48% of the African labor force. This work by women is a crucial effort in nations where the economy is usually based on agriculture.

Belying their importance to society and the economy, women have traditionally benefited from few of the structures designed to promote human capacity and ability to innovate. UNESCO estimates that only 45% of women in Africa are literate, compared to 70% of men; 70% of African women do not complete primary school, and only about 1.5% of women achieve higher education. Of all of the disciplines, science and agriculture attract the fewest women.

For example, in Ghana, women account for only 13% of university-level agriculture students and 17% of scientists.[2] By not focusing on building the capacity of women, African states miss the chance to increase the productivity of a large portion of their labor force and food production workers. The lack of female involvement in education, especially science and agriculture, means there is an enormous opportunity to tap into skills and understandings of agricultural production that could help lead to more locally appropriate farming techniques and more thorough adoption of those techniques.

Gender and Agriculture

Women play a crucial role in agriculture. Although nearly 50% of agricultural producers in sub-Saharan Africa are women, they have limited access to land, technology, financial services, education, and markets. As a result, yields on their plots tend to be lower. In fact, an FAO study estimates that granting women equal access to inputs and resources could lead to a 20%–30% increase in production.³ This is important not only for individual and family well-being but for food security overall. Only 69% of female farmers receive visits from agricultural extension agents, compared to 97% of male farmers. Of agricultural extension agents, only 7% are female. In many places, it is either culturally inappropriate or simply uncommon for male extension agents to work with female farmers, so existing extension systems miss the majority of farmers. Additionally, as the Central American case below demonstrates, having extension workers who understand the experience of local farmers is central to promoting adoption. An important component of successful adoption is including female extension workers and educators in formal and informal settings.

Women's access to technology is also significantly lower than men's. Access to new technology is critical for improving productivity. Gender gaps are most pronounced in access to machinery and tools, inputs such as improved seeds and fertilizers, and farming techniques such as sustainable pest and land management. Women tend to be at a disadvantage when it comes to using purchased inputs because they lack access to credit and suffer from less education and greater time constraints than male farmers. This inequality is most pronounced in South Asia and sub-Saharan Africa. There are specific examples of this. Thirty-nine percent of Ghanaian female farmers adopt improved crop varieties, compared to 59% of male farmers. In Kenya, Malawi, and Benin, households headed by women have lower adoption rates of improved seeds and fertilizers due to less access to land, labor, credit, and education. Burkinabé female farmers use less fertilizer per hectare than male farmers. And where farmer groups share machinery and equipment, female farmers must wait until male farmers finish plowing; as a result of the delay, their yields are lower and they are unable to complete a second planting.⁴

Unequal distribution of education is the other critical factor in the misuse of women's contributions in agricultural production. Compared to the colonial period and the situation inherited at independence, considerable gains have been realized in general and female school attendance. Still, many countries have not yet achieved even universal primary school attendance. Gender inequality is most severe in contexts where general enrollment is lower.

Furthermore, several countries had a severe setback in the early 1980s, as their enrollment rates were either stagnant or declining. The persistent economic crisis meant that the previously agreed-upon targets of universal primary enrollment in 1980 and then 2000 could not be met. Since the Dakar Conference of 2000 and the creation of the Millennium Development Goals, new targets were set for universal primary enrollment by 2015. However, at this stage, there is little doubt that most countries will not be able to reach this goal. By and large, countries with the lowest enrollment ratios from primary to higher education levels have the lowest enrollment ratios for their female populations.

African countries have shown considerable vitality in enrollment in higher education since the mid-1990s, following the lean years of the destructive structural adjustment programs. Nevertheless, African countries still have the lowest higher education enrollment in the world. Although there are a few exceptions in southern Africa (Lesotho is a unique case, where nearly three-fourths of the higher education students are females), in most African countries female enrollment is lower than that of males. Furthermore, the distribution of higher education students by discipline shows consistently lower patterns of female representation in science, technology, and engineering.

Considering African women's cultural heritage and continued central role in agriculture, it is a major paradox that their representation is so low in tracks where agricultural extension workers and other technicians and support staff and agricultural engineers are trained. Indeed, if there were any consistency between current educational systems and adequate human resource development, there would be at least gender parity in all the fields related to agriculture and trade. Yet only a few countries, such as Angola and Mozambique, have designed and implemented policies encouraging a high representation of females in science, including those fields related to agriculture. More generally, for both males and females, little effort is made in the educational system to promote interest in science in general and agriculture in particular.

It is vital to put more emphasis on involving women in agriculture and innovation, as well as helping female farmers build their capacity to increase productivity. There are several avenues to reach this goal. The first is women's training programs that focus specifically on agriculture. Another crucial avenue is emphasizing female participation in extension work—both as learners and extension agents, as discussed later in this chapter.

African Rural University (ARU) for women in the Kibaale district of western Uganda, established and incubated by the Uganda Rural Development and Training Program (URDT), is one of the first universities to focus on educating and training women to be professionals in rural development. URDT, an endogenous rural NGO, has had huge success in supporting change in the region since its founding as an extension project in 1987. Its impact has resulted in better food security, increased educational attainment, better nutrition, strong female leaders who engage in peace-building efforts, and higher incomes for families and communities across the district. One driving factor is the innovative model of community-university interaction that focuses on women and agriculture.[5]

URDT has leveraged its 30-year experience in rural development work to create programs that enhance the inherent capacities of women as leaders, nurturers, economists, agriculturalists, organizers, and health educators. The core programs train learners to gain mastery in the visionary approach, systems thinking, and principles of sustainable development. African Rural University (ARU) is the cornerstone of URDT's strategy to create a critical mass of rural transformation professionals, to advance university-level development research, and to give rural communities influence in the national development process.

The African Rural University for women provides transformational education to create effective change agents within an African development context. ARU's programming is tailored to meet locally identified needs that value existing lifestyles and traditions while allowing the adoption of new technologies and improved production. As change makers and agents, the students of ARU are trained to enable people in rural communities to envision the future they want and to design strategies to achieve their goals and aspirations. Community members are involved in the identification, development, piloting, and use of agricultural technologies and innovations. The students' four-year engagement with faculty and community members leads to sustainable food security, commercialism of agriculture, and gender empowerment.

Agriculture is one of the dominant sectors of practice in rural Africa. Unfortunately, it has remained subsistence in nature due to cultural, skill levels, resource, and governance challenges. It is against this backdrop that ARU training takes a highly entrepreneurial approach to agriculture while promoting the democratic practices around building a shared vision of transformation at a personal, household, community, national, and continental level. The ARU model for higher education enables young graduates as professionals to learn to value staying in rural areas, hence providing the necessary quality of human resources needed for transformation.

Right from the first semester, a student is allocated a practice garden to experiment with various crop technologies, to understand the agronomics as well as the economics of a given crop enterprise. Following a curriculum that is 40% practice, students continue to work hands on with local communities, initiating and monitoring modern farm technologies such as high-value crops, greenhouses, and solar projects. ARU students have been helpful in organizing community members to start a savings and credit program with the intention of acquiring solar power in their homes.

The power of this model becomes apparent when you consider the example of one ARU student who was placed in a community that had been written off by the local government as impossible to undertake any development initiatives. This ARU student has been instrumental in transforming that community into a model community within two years.

URDT also has an award-winning primary and secondary Girls' School, founded in 2000, which focuses on developing girls' abilities in a variety of areas, including agriculture, business, and leadership, along with the standard national curriculum. At URDT Girls' School, bright girls from disadvantaged backgrounds engage in a Two-Generation Approach and "Back Home" projects with their families, where they work on a project designed to improve the situation at home. Such projects include creating a community garden, building drying racks to preserve food in the dry season, and conducting hygiene education. Parents come to the school regularly to also engage in education and help the girls design the Back Home projects. School becomes both a learning experience and a productive endeavor; therefore, families are more willing to send children, including girls, to school because they see it as relevant to improving their lives.

To a family in poverty, the economic benefits of sending a girl to school are often not clear. Using the Back Home projects for income generation, one of the remarkable URDT students was able to keep herself and siblings in school after she was orphaned. African Rural University provides a logical next step for URDT Girls' School graduates and other women who want to be leaders in improving rural life.

Another graduate of the URDT Girls' School went on to become an exemplary ARU student. Building on her experience of Back Home projects as well as the fieldwork assignments at ARU, she was able to start a permaculture project for youth in her home village who had dropped out of school. The school, now in its third year, has 280 pupils and it is undergoing national registration process.

African Rural University students are given a one-month residency in homes of URDT Girls' School students, where they engage the households and their immediate communities, evaluate the visions that they have developed, and assess strategies of obtaining the visions. They support the development and strengthening of associations of parents with students from URDT Girls' School. The associations have been a source of promoting enterprise development in such communities. This partnership demonstrates a continuum of integrated rural development between the primary, secondary, and university education.

URDT focuses on agriculture and on having a curriculum that is relevant for the communities' needs. There is an experimental farm where local people can learn and help develop new agricultural techniques, as well as a Vocational Skills Institute to work with local artisans, farmers, and businessmen who have not had access to traditional schooling. There is an innovative community radio program designed to share information with the broader community. URDT also runs an Appropriate and Applied Technology program that allows people from the community to interact with international experts and scientists to develop new methods and tools to improve their lives and agricultural productivity.

Governments can draw on this model to create effective learning institutions to support agriculture, and particularly women's and communities' involvement in it. The three key lessons of the model are to make sure that the school is working with and giving back to the community by focusing on its needs, which are often based around agriculture; to create a holistic program that sees how the community and the institution can work together on many interventions—technology, agriculture, market infrastructure, and education—to improve production and the standard of living; and to focus on women and girls as a driving force behind agriculture and community change, benefiting the whole society.

The crucial unifying factor is to integrate education at all levels, and the research processes of higher education in particular, back into the community. This allows the universities to produce technologies relevant to rural communities' needs and builds trust among the research, education, and farming communities.

Governments can draw on this model to create effective learning institutions to support agriculture, and particularly women's and communities' involvement in it. The three key lessons of the model are to make sure that the school is working with and giving back to the community by focusing on its needs, which are often based around agriculture; to create a holistic program that sees how the community and the institution can work together on many interventions—technology, agriculture, market infrastructure, and education—to improve production and the standard of living; and to focus on women and girls as a driving force behind agriculture and community change, benefiting the whole society.

The crucial unifying factor is to integrate education at all levels, and the research processes of higher education in particular, back into the community. This allows the universities to produce technologies relevant to rural communities' needs and builds trust among the research, education, and farming communities.

Community-Based Agricultural Education

Uganda is not alone in adopting this model. The government of Ghana established the University for Development Studies (UDS) in the northern region in 1992. The aim of the university is to bring academic work to support community development in northern Ghana (Brong-Ahafo, Northern, Upper East, and Upper West Regions). The university includes agricultural sciences, medicine and health sciences, applied sciences, integrated development studies, and interdisciplinary research. It relies on the resources available in the region.

UDS seeks to make tertiary education and research directly relevant to communities, especially in rural areas. It is the only university in Ghana required by law to break from tradition and become innovative in its mission. It is a multi-campus institution, located throughout northern Ghana—a region affected by serious population pressure and hence is vulnerable to ecological degradation. The region is the poorest in Ghana, with a relatively high child malnutrition rate. The university's philosophy, therefore, is to promote the study of subjects that will help address human welfare improvement.

The pedagogical approach emphasizes practice-oriented, community-based, problem-solving, gender-sensitive, and interactive learning. It aims to address local socioeconomic imbalances through focused education, research, and service. The curricula stress community involvement and community dialogue, extension, and practical tools of inquiry.

Students are required to internalize the importance of local knowledge and to find effective ways of combining it with science. The curricula also include participatory rural appraisal, participatory technology development, and communication methodologies that seek to strengthen the involvement of the poor in development efforts.

An important component of the emphasis on addressing sustainable development is the third trimester practical field program. The university believes that the most feasible and sustainable way of tackling underdevelopment is to start with what the people already know and understand. This acknowledges the value of indigenous knowledge. The field program brings science to bear on indigenous knowledge from the outset.

Under this program, the third trimester of the academic calendar, lasting eight weeks, is exclusively for fieldwork.

Students live and work in rural communities. Along with the people of the community, they identify development goals and opportunities and design ways of attaining them. The university coordinates with governmental agencies and NGOs in the communities for shared learning in the development process. The field exposure helps students build up ideas about development and helps them reach beyond theory. The impact of this innovative training approach is already apparent, with the majority of UDS graduates working in rural communities.

Early Agricultural Education

In order for children to engage in agriculture and understand it as a part of their life where they can build and develop skills and abilities to improve their future, it is necessary to continue their exposure to agricultural techniques and skills throughout their education. Equally important is the need to adapt the educational system to reflect changes in the agricultural sector.[6] Many rural African children will have been to the family farm or garden, and will have done some small work in the field, before they ever arrive at school. Children go with their mothers into the field from a very young age and so are likely to be familiar with local crops and the importance of the natural world and agriculture in their lives. Schools can capitalize on this early familiarity as a way to keep children engaged in the learning process and to build on skills that will help them increase their production and improve their lives for the future.

School Gardens

One model to achieve early engagement is by having a school garden. Schools all over the world, from the United States and the United Kingdom, to Costa Rica and Ecuador, to South Africa and Kenya, use school gardens in various guises to educate their students about a set of life skills that goes beyond the classroom. School gardens come in many forms, from a plot of land in the school courtyard, to the children visiting and working in a broader community garden, to planting crops in a sack, a tire, or some other vessel. These gardens can use as many or as few resources as the community has to devote to them. The sack gardens especially require very few resources and can be cultivated in schools with little arable land and in urban areas. Students can also bring the sack garden model back home to their families to improve the family's income and nutrition.

Labor in the school garden should certainly not replace all other activities at the school, but should serve as a complement to the other curriculum; it can provide a place where students learn important skills and feel that they are productive members of their community. Children who participate in school gardens learn not only about growing plants, food, and trees—and the agricultural techniques that go along with this—but also about nutrition, food preparation, responsibility, teamwork, and leadership. As students get older, they can also use the garden and the produce it generates as a way to learn about marketing, economics, infrastructure needs, and organizing a business. Many schools have student associations that sell their produce in local markets to learn about business and generate income.

School gardens have the added benefit of showing communities that the government recognizes agriculture as an important aspect of society and not as a secondary endeavor. Schools that provide education in gardening often overcome parents' reluctance to send children to school, as they teach a set of skills that the parents recognize as being important for the community—and parents do not see schooling as the loss of the child's potential labor at home. A government can increase this impact by involving the community in educational programs and curriculum decisions. Promoting buy-in from the community for the entire educational process encourages families to enroll more students and allows children to learn important skills.

Also, by valuing agriculture and enabling more productive work in the community, school gardens decrease the incentive for large migrations to urban areas. This also calms many parents' fears that a child who goes to school will leave home and will not continue to work on the family farm. This emphasis on agriculture benefits both children and parents, by giving them access to a formal education and a way to increase agricultural productivity.

Semiformal Schooling

Another model that can work to encourage children and young people to learn agriculture is a semiformal schooling model. Here, children spend part of the day at school learning math, literacy, and traditional subjects, and part of the school day working in a field or garden. This second part of the day is a chance to generate some income for families as well as to learn new agricultural and marketing techniques. Generally, these kinds of programs are for older children who have never gone to primary school; they are taught in local languages instead of the official English or French of many formal school systems. This model can be adapted for adults as well to encourage literacy and the development and adoption of new agricultural techniques. In South Africa, this model is often referred to as a Junior Farmer's Field School, to get young people involved early in the experiential process of learning and creating new agricultural techniques.

School gardens, the inclusion of agriculture in the formal curriculum, and technical training models are all ways to promote children's experience with agriculture and help them develop the skills they need to improve their livelihoods into adulthood. These models place value on agriculture, the local community, and the process of experience to encourage children to learn new skills and engage in the natural world in a productive way.

Experiential Learning

There are several examples of how farmers can play a role in experimenting with new innovations, making them feel a sense of ownership of related tools and increasing the chance that other farmers will use the techniques. These examples also show how innovations work in the field and what changes are needed for better results.

Nonformal educational systems are crucial for reaching the population that is past the age for traditional primary schools and for encouraging local adoption of new techniques. Even if revolutionary new technologies exist at the research level, they can improve economies only if farmers use them, so getting information into the hands of local farmers, especially women, is vital to the success of research endeavors and should be part of any plan for agricultural growth.

Two of the persistent obstacles to the adoption of new peanut varieties are the difficulty of obtaining the seeds and the reluctance to use new seeds without being sure how they will grow compared to the traditional variety. Farmers want many of the benefits that new seed varieties can bring—they typically prefer high-yield, high market value, pest-resistant, and high oil-content varieties—but often they cannot get the seeds or they are afraid the new seeds will fail. Without some guarantee that the new seeds will work, farmers are often unwilling to risk planting them, even if they are readily available, and these farmers are certainly unwilling to make the substantial investment of time and capital that is usually required to seek out and acquire new seed varieties. Not many rural farmers have the resources to go to the capital city and purchase experimental seed varieties from a research institute, and the risk of an unknown variety is often too high for a family to take.

One way of addressing this challenge is to give trial seed packages to pilot farmers or members of the local farmers' association to try on a portion of their land or on a test plot. This is a variation of the early adopters' model, which searches for members of a community who are willing and able to take some risk, and who then spread an idea to their peers. This strategy addresses both difficulties, since it allows for a trial with minimal risk, as well as a local source for new seed. Once the pilot farmer or association members grow the new variety of seed, they can sell it to their neighbors.

The International Crops Research Institute for the Semi-Arid Tropics (ICRISAT), in partnership with the Common Fund for Commodities, has developed a trial package for new varieties of peanut in the Sahel (tested in Mali, Niger, and Nigeria) and disseminates it through pilot farmers and farmers' associations. These farming associations are often women's associations, since women traditionally need cash crops to be able to meet their families' economic needs but have even less access to improved seed varieties than men. In all of the countries, ICRISAT provided 17 kilograms of new seed varieties to their pilot farmers, as well as training in field management techniques that maximize the yield for their crop. The project's agents then asked local farmers to help distribute the new seed varieties through the members of their associations.[7]

Although the management techniques were imperfectly applied, and there is a cost associated with the new varieties and techniques, farmers using the new varieties experienced substantial returns on their investment. New varieties were 97% more profitable than traditional ones, so farmers earned almost twice as much on their investment as they would have with the types of seed already in use.

This is a story that has repeated itself often over the trials. ICRISAT has learned that the people most likely to adopt new peanut varieties—and who therefore make good pilot farmers—are those who are slightly younger, have smaller family sizes, and have relatively more access to resources, such as labor and land. These are the people who can afford to take a risk at the beginning, and when that risk pays off, they serve as a model for the other farmers in their community. They will then also serve as a source of local seed, which is very important, since farmers are most likely to use either their own seed stocks or stocks available from local markets.

Through this model, small investments can spread the use of modern seed varieties that have much higher yields and are more profitable to sell. These higher yields and profits ensure food security, including much-needed protein in rural diets, while improving the quality of life for the farmer.

As mentioned earlier, sending extension workers—either from governments or NGOs—into the field is a common practice that can be more or less effective, depending on who the extension agents are and how they handle the situation in the villages. Extension agents who are the peers of local villagers and who practice the lessons they teach in a way that the other farmers can observe are usually the most effective.

Many countries in Africa have a variety of ethnic groups and regional subgroups who have different habits, speak different languages, and have different resources. The further removed an extension agent is from the population with which he or she works—by barriers of language, socioeconomic status, gender, education, or tradition—the more difficult it is to convince people to adopt the technique. There is a tendency for people to decide that the idea is appropriate for someone like the extension agent, but not someone like themselves, even if they think that the idea is a good one. The comment, "That may be how that group does it, but it could never work in our village," is a common one for formal educators who come from a city or a different population subgroup. However, if the teacher is a peer, it is harder to make the distinction between their success and the potential success of each village farmer.

Governments can use the peer educator or farmer-to-farmer method to help spread information and new agricultural innovations across their entire rural population. By funding a few formal extension workers who train and help support a large network of peer educators, a government can reach most or all of the rural population, even if the groups are segregated by language, ethnicity, geography, or traditional farming techniques. Thus, a relatively small investment can have huge impacts on a country's agricultural processes and therefore on food security and the national economy.

Farmer Field Schools (FFS) provide a way for communities to test a new technique and adapt it to their own specific needs. Many agricultural technologies need to be adapted to local contexts once they leave the lab to ensure that they are practical for farmers and that people can adopt them into their current agricultural practices. The FFS also allow for easier dissemination of new information because peers, as opposed to outsiders, are the teachers. This model also develops community ownership by encouraging local participation in new processes and leads to better adoption among participants.[8]

Local farmers participating in FFS are often selected through local leadership structures or village farming associations. They plant one plot using the techniques that they currently use and a second plot with the new technology. At the end of the growing season, the farmers then come together to compare the costs, revenues, and profits between the old way and the new technology. In this way, farmers can see what works for them and can adapt the new method as seems appropriate during the growing season. Farmers also become invested in the process and have reason to believe that it will work for them.

Any organization—private or public—can start a Farmer Field School. The resources needed are access to the new technology, be it a seed variety, a new fertilizer, or a new irrigation technique; a few extension agents to train a cadre of local farmers to spread the innovation; and a few follow-up visits to monitor the process and help villages interpret the results. These results should then move up to the national level to inform state policy and research. The following is an example of how an FFS can be used to address a specific problem.

Striga, often called witchweed, is a plant that grows in millet, sorghum, and other cereal fields across West Africa and causes myriad problems.[9] It can reduce crop yields between 5% and 80%, reduce soil fertility, and erode soil, all of which decrease the durability and profitability of rainy season–based agricultural systems. A single weed can produce more than 200,000 seeds, which remain viable in the soil for up to 10 years, making the plant very hard to eradicate. There are places where *Striga* infestation means that farmers lose money on every cereal crop they plant and are unable to feed their families or earn a living.

Nevertheless, there is a solution. In 2007, the International Crop Research Institute for the Semi-Arid Tropics, the International Fund for Agricultural Development, and several European agricultural research organizations partnered with the Tominion Farmers' Union in Mali to implement a project that uses an integrated management system combining intercropping with beans or peanuts, reduced numbers of seeds per hole planted, and periodic weeding to control *Striga*. Using the FFS model, a few agricultural experts trained 75 local farmers to train their peers in integrated management techniques. These farmers then trained 300 others, and implemented the test plot procedure for their areas.

The results are impressive. *Striga* plants decreased, crop yields and profits increased, and many farmers decided to implement the process in their own fields. Farmers discovered that it was necessary for them to conduct three cycles of weeding, rather than the two that the project originally recommended. This change has been formally adopted into the integrated management system. With these three weeding cycles, the incidence of *Striga* in the field went to zero in the test plots. Profits per hectare increased from $47 to $276, an improvement of nearly 500%. In some cases, villages went from a loss to a profit on their fields. The return on investment more than tripled, so while there was a slightly increased cost of the new methods, they more than paid for themselves. Many of the farmers involved in the 2007 study used the new methods in their own fields in 2008, and spread the message about the new techniques to their neighbors. This encouraged an enabling environment for the adoption of new technologies.

Another model is that of radio education—mentioned in the URDT case above—where extension education sessions are recorded in the appropriate local languages and are broadcast periodically on the radio. For many rural communities with low access to television and low literacy, this can be a crucial way to spread information to local farmers, especially if done in conjunction with another model, like the technical training or extension models that allow farmers to ask follow-up questions.

Innovation in Higher Agricultural Education

America's land grant colleges pioneered agricultural growth by combining research, education, and extension services. The preeminent role of universities as vehicles of community development is reflected in the US land-grant system.[10] The system not only played a key role in transforming rural America but also offered the world a new model for bringing knowledge to support community development. This model has found expression in a diversity of institutional innovations around the world. While the land-grant model is largely associated with agriculture, its adaptation to industry is less recognized. Universities such as the Massachusetts Institute of Technology (MIT) and parts of Stanford University owe their heritage to the land-grant system.[11] The drift of the land-grant model to other sectors is not limited to the United States. The central mission of using higher education to stimulate community development is practiced around the world in a variety of forms.

There are three models for entrepreneurial education in Brazil that have advanced to different stages of creating an "entrepreneurial university."[12] The Pontifical Catholic University of Rio de Janeiro, the Federal University of Itajubá, and the Federal University of Minas Gerais have all started to include entrepreneurship in the educational experience of their students. This experience often complements and coordinates with private-sector initiatives, and in some cases companies fund parts of the curriculum. The interaction between academia, government, and industry allows for a broader approach and for a shifting of program goals.

The lessons from these three schools are that flexibility in curricula and openness to partnering with other organizations—especially industry—allow universities to develop successful entrepreneurship programs that provide employment opportunities for their students, as well as a chance to experience the culture of starting a business. The stimuli that lead universities to these activities might be an external change—lack of funding from the government—or an internal decision to shift focus. An institution that is more flexible, whose staff supports the change in a more unified way, is more likely to make the change toward becoming an "entrepreneurial university," which allows students to focus on not just having business know-how and the ability to work for or with large companies, but also on how to create jobs and opportunities for themselves and their peers. Universities must have the autonomy and the flexibility to adopt these programs, as well as the ability to build networks with local actors. Ultimately, this will contribute to the nation's development.

African countries would be better served by looking critically at these variants and adapting them to their conditions. These institutional adaptations often experience opposition from advocates of incumbent university models. Arguments against the model tend to focus on the claim that universities that devote their time to practical work are not academic enough. As a result, a hierarchy exists that places such institutions either at the lower end of the academic ladder or simply dismisses them as vocational colleges.

One option that might work well in African countries is a corporate college approach. Pilot projects are being implemented in Zambia and Ghana. The idea is to create training centers that are located on working farms. AGCO, for example, committed to spending US$100 million and began by establishing a training center on a "model farm" just outside Lusaka, Zambia, in June 2012. They took over a previously existing farm comprising 150 hectares. AGCO is one of the few Grow Africa companies that made significant progress on its commitments. During Phase One, they leased the land, established a demonstration farm that primarily grew maize and soya, and ran numerous training courses on operating and maintaining large machinery. Trainees included new clients as well as local farmers. During Phase Two, AGCO brought in new partners to help expand operations and infrastructure on the farm (including building a lab for soil testing and implementing an irrigation system) as well as conducting community development projects, such as teaching conservation farming techniques and establishing a 4-H project with the local schools.

In 2009, Africa Atlantic Franchise Farms (AAFF) established a working commercial farm in the Afram Plains region of Ghana with the intent of creating a system of agribusiness research and training centers, which they call African Agribusiness Knowledge Centers (AAKC), alongside AAFF and other farms in other countries. AAFF secured a 50-year land lease title for agricultural development. When it began operations, AAFF farmed a smaller plot of land to establish best practices. Using a pivot irrigation system, it primarily grows maize for local production and distribution. The farm recently completed a rigorous environmental and social impact assessment following World Bank standards. The goal of AAKC is to educate and equip farmers to become agricultural entrepreneurs. It instructs farmers on sustainable farming techniques and farm management practices.

The land-grant model is being reinvented around the world to address such challenges. One of the most pioneering examples in curriculum reform is EARTH University in Costa Rica, which stands out as one of the first sustainable development universities in the world.[13] It was created in 1990 through a US$100 million endowment provided by the US Agency for International Development (USAID) and the W. K. Kellogg Foundation. Its curriculum is designed to match the realities of agribusiness.[14] The university dedicates itself to producing a new generation of agents of change who focus on creating enterprises rather than seeking jobs.

EARTH University emerged in a context that mirrors today's Africa: economic stagnation, high unemployment, ecological decay, and armed conflict. Inspired by the need for new attitudes and paradigms, EARTH University is a nonprofit, private, international university dedicated to sustainable agricultural education in the tropics. It was launched as a joint effort between the private and public sectors in the United States and Costa Rica. The Kellogg Foundation provided the original grant for a feasibility study at the request of a group of Costa Rican visionaries. Based on the study, USAID provided the initial funding for the institution. The original mission of the university was to train leaders to contribute to the sustainable development of the humid tropics and to build a prosperous and just society. Located in the Atlantic lowlands of Costa Rica, EARTH University admits about 110 students a year and has a total student population of about 400 from 24 countries (mainly in Latin America and the Caribbean) and faculty from 22 countries. Through its endowment, the university provides all students with 50% of the cost of tuition, room, and board.

In addition, the university provides scholarships to promising young people of limited resources from remote and marginalized regions. Nearly 80% of the students receive full or partial scholarship support. All students live on campus for four intensive years.

EARTH University has developed an innovative, learner-centered, and experiential academic program that includes direct interaction with the farming community.[15] Its educational process stresses the development of attitudes necessary for graduates to become effective agents of change. They learn to lead, identify with the community, care for the environment, and be entrepreneurial. They are committed to lifelong learning. There are four activities in particular within the curriculum that embody EARTH University's experiential approach to learning.

Learning from Work Experience and Community Service

The first is the Work Experience activity, which is taken by all first-, second-, and third-year students and continues in the fourth year as the Professional Experience course. In the first and second years, students work on specific projects on EARTH University's 3,300-hectare farm. In the first year, the work is largely a routine activity and the experience centers on the acquisition of basic skills, work habits, and general knowledge and familiarity with production. In the second year, the focus changes to management strategies for these same activities. Work Experience is later replaced with Professional Experience. In this course, students identify work sites or activities on campus that correspond with their career goals. Students are responsible for contacting the supervisors of the campus operations, requesting an interview, and soliciting "employment." Upon agreement, supervisors and students develop a joint work plan that the student implements, dedicating a minimum of 10 hours per week to the "job." The second activity is an extension of the Work Experience course. Here third-year students work on an individual basis with small, local producers on their farms. They also come together in small groups under the Community Outreach program that is integral to the learning system. Community outreach is used to develop critical professional skills in students, while at the same time helping to improve the quality of life in nearby rural communities. The third-year internship program emphasizes experiential learning. The 15-week internship is required for all students in the third trimester of their third year of study. It is an opportunity for them to put into practice all they have learned during their first three years of study. For many of them, it is also a chance to make connections that may lead to employment after graduation. The international character of the institution allows many students the opportunity to follow their interests, even when they lead to internship destinations other than in their home country.

Sharpening Entrepreneurial Skills

The fourth activity is the Entrepreneurial Projects Program. EARTH University's program promotes the participation of its graduates in the private sector as a critical means by which the institution can achieve its mission of contributing to the sustainable development. The development of small and medium-sized enterprises (SMEs) is a powerful way to create new employment and improve income distribution in rural communities. For this reason, the university stresses the development of an entrepreneurial spirit and skills. Courses in business administration and economics, combined with practical experience, prepare the students to engage in business ventures upon graduation.

This course provides students the opportunity to develop a business venture from beginning to end during their first three years at EARTH University. Small groups of four to six students from different countries decide on a relevant business activity. They conduct feasibility studies (using financial, social, and environmental criteria), borrow money from the university, and implement the venture. This includes marketing and selling the final product. After repaying their loan, with interest, the group shares the profits. This entrepreneurial focus has permeated all aspects of the university's operations and has prepared students to become job creators and agents of change rather than job seekers. The university also manages its own profitable agribusiness, which has resulted in strong relationships with the private sector. When the university acquired its campus, it decided to continue operating the commercial banana farm located on the property. Upon taking over the farm, the university implemented a series of measures designed to promote more environmentally sound and socially responsible production approaches.

Going Global

EARTH University has internationalized its operations. It signed an agreement with US-based Whole Foods Market to be the sole distributor of bananas in their stores. The university also sells other agricultural products to the US market, among others. This helps to generate new income for the university and for small farmers while providing an invaluable educational opportunity for the students and faculty, as well as contributing to the EARTH University scholarship fund. The university uses part of the income to fund sustainable and organic banana and pineapple production research.

Over the years the university has worked closely with African institutions and leaders to share its experiences and currently has students from 14 African countries. After many years of sharing experiences with colleagues from a number of African universities, hosting delegations and offering workshops in several African countries, EARTH has become widely recognized as an innovative institution whose educational model has great relevance for the African continent. Currently there are many ongoing conversations regarding the possibility of establishing similar institutions there. The case of EARTH University is one of many examples around the world involving major collaborative efforts between the United States and East African countries to bring scientific and technical knowledge to improve welfare through institutional innovations. Such experiences, and those of US land-grant universities, offer a rich fund of knowledge that should be harnessed for Africa's agricultural development and economic growth.

Such models reveal ways to focus agricultural training as a means of improving practical farming activities. Ministries of sustainable agriculture and farming enterprises in East African countries should be encouraged to create entrepreneurial universities, polytechnics, and high schools that address agricultural challenges. Such colleges could link up with counterparts in developed or emerging economies, as well as institutions providing venture capital, and start to serve as incubators of rural enterprises. Establishing such colleges will require reforming the curriculum, improving pedagogy, and granting greater management autonomy. They should be guided by the curiosity, creativity, and risk-taking inclination of farmers.

National Agricultural Research Universities

Perhaps the best starting point for improving agricultural education in sub-Saharan Africa is the national agricultural research institute (NARI). It is notable that most African universities do not specifically train agriculture students to work on farms in the same way that medical schools train students to work in hospitals. Part of the problem arises from the traditional separation between research and teaching, with research carried out in national research institutes and teaching in universities. There is little connection between the two in many African countries.

There are two main reasons for this separation and the associated fragmentation. Africa established colonial research institutes before it created universities. The main function of the research institutes was to serve colonial agricultural objectives, rather than building local scientific and technological capabilities or fostering local entrepreneurship.

The first generation of African universities was designed to prepare young Africans for public service and as a result focused largely on the social sciences and humanities. By the time universities were being established, the European tradition of separating research from education was already in place. This separation found expression in distinct laws as well as in ministries. This approach, also expressed in ministerial separation, is more evident in former British colonies than in Francophone countries.

The second reason for the separation is legislative continuity and emulation. African countries continued the same tradition partly because their economic structures did not create much demand for locally generated knowledge, except in fields such as agriculture. African countries continued to reproduce the structure, despite the fact that it did not appear to reflect local realities. For example, much of the research cooperation among foreign universities is conducted through national research institutes. This hampers the ability of African countries to foster stronger international university-to-university partnerships.

The fragmentation was worsened by two additional factors. First, agricultural extension services that used to exist in agricultural ministries collapsed in the 1980s largely because of cutbacks in public expenditure. Second, there are no major efforts aimed at commercializing local research results. The absence of extension support and lack of mechanisms that foster commercialization have left NARIs considerably isolated, and have undermined their ability to promote innovation.

Policy Lessons

The challenges facing African agriculture will require fundamental changes in the way universities train their students so that agricultural education can contribute directly to the agricultural sector. The NARIs in Africa operate a large number of research programs that provide a strong basis for building new initiatives aimed at upgrading their innovation capabilities. In effect, what is needed is to strengthen the educational, commercialization, and extension functions of the NARIs.

A number of critical measures are needed at the regional and national levels to achieve this goal.[16] The first should be to rationalize existing agricultural institutions by designating some universities as hubs in key agricultural clusters. More specifically, clustering these functions would result in dedicated research universities whose curriculum would be modeled along full value chains of specific commodities. For example, innovation universities located in proximity to coffee production sites should develop expertise in the entire value chain of the coffee industry. This is particularly important in the face of climate change. The coffee berry borer (*Hypothenemus hampei*), for example, is thriving in the increased temperatures in East Africa, causing damage to crops. Coffea-arabica regions in Ethiopia, parts of Uganda, Kenya, and most of Rwanda and Burundi will likely be affected in the future. One strategy is to plant shade trees to adapt to rising temperatures, but this will not address the root of the problem. In fact, "there is a pressing need to fill existing knowledge gaps in the coffee industry, and to develop science-based adaptation strategies to mitigate the impacts of climate change on coffee production."[17] Adapting agriculture to climate change will require the intensified application of science, technology, and innovation, focusing on specific crops. A similar situation is occurring in the cacao-producing regions of West Africa, where production is falling far short of demand. Establishing a crop-specific, research-oriented, entrepreneurial university will help to consolidate research efforts toward sustainable solutions for cacao tree crops in a changing climate. The university's curriculum will include teaching young rural entrepreneurs and mentoring them in agribusiness with a paradigm of "crops as business." This could be applied to other crops, as well as livestock and fisheries. Such universities could be designed around existing national research institutes that would acquire a training function as part of a regional rationalization effort. Such dedicated universities would not have monopoly over specific crops but should serve as opportunities for learning how to connect higher education to the productive sector.

Internally, the universities should redefine their academic foci to adjust to the changes facing Africa. This can be better done through continuous interaction among universities, farmers, businesses, government, and civil society organizations. Governance systems that allow for such continuous feedback to universities will need to be put in place.

The reform process must include specific measures. First, universities need a clear vision and strategic planning for training future agricultural leaders with a focus on practical applications.[18] Such plans should include comprehensive road maps on how to best recruit, retain, and prepare future graduates. These students should be prepared in partnership with key stakeholders.

Second, universities need to improve their curricula to make them relevant to the communities in which they are located. More important, they should serve as critical hubs in local innovation systems or clusters. The community focus, however, will not automatically result in local benefits without committed leadership and linkages with local sources of funding.[19] The decision by Moi University in western Kenya to acquire an abandoned textile mill and revive it for teaching purposes is an example of such an opportunity.

Third, universities should give students more opportunities to gain experience outside the classroom. This can be done through traditional internships and research activities. But the teaching method could also be adjusted so that it is experiential and capable of imparting direct skills. More important, such training should also include the acquisition of entrepreneurial skills.

Fourth, continuous faculty training and research are critical for maintaining high academic standards. Universities should invest more in undergraduate agricultural educators to promote effective research and teaching and to design new courses.

Fifth, in addition to degree courses, universities for agricultural innovation will also need to extend their reach into the sphere of vocational training. This can be done directly through various programs such as "farmer schools" or in conjunction with high schools. The link with high schools and other educational institutions is particularly important considering Africa's demographic structure. In most parts of the continent, the majority of the population is in school, which makes educational institutions an integral part of the community.

Sixth, one way to facilitate the transfer of knowledge from universities to farming communities is through internships and community service. These activities should be structured so that they are part of the academic calendar. They would serve two main purposes. The first would be to transfer knowledge from universities to farmers. Second, returning students would bring back to the university feedback and lessons that could be used to adjust the curriculum, pedagogy, and interactions with farmers.

Seventh, one of the main teaching missions of universities for innovation is to translate ideas into goods and services through enterprise development. Training young people to learn how to create enterprises should therefore be part of the mission of such universities. This can be done in partnership with financial institutions such as banks, cooperatives, and microfinance organizations. Such activities may also lay the foundation for the emergence of rural-based angel funding or venture capital facilities. Similarly, sources of support such as rural development funds could be redirected to help translate ideas from such universities into new enterprises.

Eighth, continuous faculty training and research are critical for maintaining high academic standards. The new universities should invest more in undergraduate agricultural educators to promote effective research and teaching and to design new courses. Researchers at NARIs would only need minimum training to acquire the necessary pedagogical skills. In fact, many of them are involved in extensive field training activities, and so they already teach without having the title. Additional support to the NARIs can be provided by education departments in existing universities. Where needed, teacher training institutes could create special courses aimed at offering training in experiential pedagogy.

Finally, it is important to establish partnerships among various institutions to support and develop joint programs. These partnerships should pursue horizontal relationships and open networking to generate more synergy and collaboration, encourage sharing of resources, and foster the exchange of students and faculty. This can be done through regional exchanges that involve the sharing of research facilities and other infrastructure.

Providing tangible rewards and incentives to teachers for exemplary teaching raises the profile of teaching and improves education. In addition, establishing closer connections and mutually beneficial relationships among all stakeholders (academia and industry, including private and public institutions, companies, and sectors) should generate further opportunities for everyone.

Creating New Agricultural Universities

There are several ways to bring about the changes described above. One method is to identify concept champions. Agriculture-related constituencies of ministers and other leaders already include people who can serve as champions and advocates for upgrading NARIs into new universities for agricultural innovation. The concept champion will be essential in advancing the ideas at the national, regional, and international levels. Champions will take responsibility for exploring the political feasibility of translating the ideas laid out in this chapter into practical action. Much of their work will involve seeking broad political support at the national and regional levels.

Second, policy and legislative reform are needed to create universities for agricultural innovation. The policy framework may already exist, as they tend to follow guidelines such as CAADP that stress the importance of investing in agricultural research. However, new legal instruments may be needed to foster the creation of new research-oriented universities. This can be done via amendments to existing laws on higher education, science and technology, research, or agriculture. Alternatively, new laws may need to be adopted that allow the creation of a separate regime that can be managed by ministries responsible for agriculture in cooperation with higher education authorities. The key element of such laws and regulations would to grant sufficient autonomy to the new institutions while fostering excellence in research and practice. Policies and laws for such universities should be written in an inclusive way so that other institutions—whether private or public—that meet the established criteria can be designated as universities for innovation.

Third, building innovation management capacity will be critical for the creation and implementation of these new universities. This will require a cadre of people with expertise in innovation management. This can be achieved through executive education offered to high-level leaders responsible for policy promotion, as well as the ultimate implementation of an agricultural innovation system. In the long run, such courses should be part of the curriculum of the new universities and should be required for those seeking to work as innovation managers.

Fourth, one of the roles of the concept champions identified above will be to pilot the idea at the national level via new projects. The purpose of the pilot initiatives will be to create a basis for learning about how best to advance the idea of universities of agricultural innovation. The pilots will be carefully chosen to maximize the chances of success and not necessarily to determine the viability of the idea. The lessons learned from the execution of the pilots will be regularly shared by African countries.

Fifth, financing is probably one of the most contentious issues in Africa's history of research and higher education. The perceived high cost of running institutions of higher learning has contributed to the dominant focus on primary education. This policy, however, has prevented leaders from exploring avenues for supporting higher technical education. Creating incentives for domestic mobilization of financial resources is essential for leveraging external support. There is a wealth of knowledge from around the world on how to finance innovation, which can be leveraged to help African countries identify the diversity of available approaches. These include public as well as private funding. A comprehensive review of known options needs to be undertaken as a matter of urgency.

Sixth, it is important to establish regional and international partnerships among various institutions to support and develop joint programs. These partnerships should pursue horizontal relationships and open networking to generate more synergy and collaboration, encourage sharing of resources, and foster the exchange of students and faculty. This can be accomplished through regional exchanges that involve the sharing of research facilities and other infrastructure. Such collaboration could be extended to include international partners through mechanisms such as the OpenCourseware Consortium, a free and open digital publication of educational materials organized as courses. The consortium includes open educational content from 200 higher education institutions and associated organizations. Its mission is to advance education and empower people worldwide through open courseware. The advent of broadband Internet through investments in fiber-optic cables offers additional opportunities for the new universities to become part of the global knowledge ecology. Many universities around the world are offering online courses and are using Internet connectivity to extend their reach to the developing world. Governments and private enterprises can help strengthen these linkages by facilitating access to broadband facilities.

Finally, the tasks laid out above will require dedication, courage, and commitment that should be recognized through agricultural innovation prizes for outstanding contributions to strengthening agricultural research, teaching, commercialization, and extension work.

Lifelong Learning Through the Private Sector

The roles of the private and public sector in lifelong learning opportunities are illustrated by the case of Peru's relatively high-tech asparagus industry.[20] Both public and private programs offer industry-specific training for employees and build on the skills that many workers get from experience in the formal education sector. Those working at the managerial level tend to receive training from La Molina—the national agricultural university. There is a tension between private- and public-sector training, as hiring managers tend to perceive graduates of private education as being of higher quality, although the public sector is able to produce more graduates and therefore better meet the industry demand for workers. Ultimately, the best arrangement is some combination of public- and private-sector education and training, as Peru has high secondary and tertiary school enrollment compared to many other Latin American countries.

Asparagus exporting requires a high level of skill because of the need to keep the asparagus under controlled conditions and package it in appropriate weights. The success of this industry relies upon investment in long-term learning for employees. There is a great emphasis on on-the-job training, whereby employees learn a specific set of job-related skills. In addition, there are both private and public vocational training programs for adults. Employers give consistently better reviews to those workers who receive on-the-job training or who complete the private-sector training programs than to those who graduate from government-run programs. Students are willing to pay for private training because the curriculum and schedule are more flexible, and they allow the students to continue their employment, in contrast to the more rigid structure in public institutions. These private institutions also generally include an internship—which serves as both student training and a relatively low-cost way for employers to recruit skilled students.

Nevertheless, these programs are not without problems; they produce fewer students than the industry needs, and they rely on employees having at least primary or some secondary education, largely as a result of Peru's relatively high levels of enrollment in secondary schools.

A high proportion of managers graduate from La Molina with degrees in agronomy or engineering. La Molina not only trains many of the skilled workers in management and agronomic skills, but it also conducts much of the research that the industry uses to have its crops meet international export standards. La Molina also conducts technology transfer with countries like the United States and Israel to adapt new techniques to local realities.

There are several training models to help farmers and plant workers acquire the skills they need. First are the private models of on-the-job training, which range from informal mentoring in the first two weeks of work to Frio Aéreo's (a consortium of 10 partners that is concerned with managing the cold chain) formalized internship program and weekly training sessions during the slower seasons. Second are private universities, such as Universidad Privada Antenor Orrego, that train technicians and managers; there are also public institutions with similar goals, whose graduates tend to be less valued by employers. Additionally, there is a public-sector youth training program that aims to help young graduates become successful agricultural entrepreneurs. Finally, there is El Centro de Transferencia de Tecnologia a Universitarios, which uses holistic approaches to develop agricultural entrepreneurs, either by giving students plots of land that they must pay for over several years, or by working with small farmers who already own land. The model that works with smallhold-ers requires an investment of about US$33,000 per farmer.

Private-sector initiatives have so far been more successful at training older workers in the necessary techniques. How-ever, much of the system depends on workers having initial basic public education, as well as the managerial expertise and public goods that La Molina provides. Successful training for high-skill industries requires a combination of private-sector initiatives and a solid foundation of public education and research.

Conclusion

The current gaps in educational achievement and the lack of infrastructure in many African school systems are an op-portunity for governments to adopt more community-driven models that prioritize education in a holistic way that will improve community involvement, child achievement, ag-ricultural production, and the standard of living for rural populations. Acknowledging that agriculture is both a valued traditional lifestyle and a huge potential driver of economic growth, and changing educational programming to respect these goals, will go a long way toward encouraging basic edu-cation and improving people's lives.

No new agricultural technology, however cutting-edge and effective, can improve the situation if people are unable to access it and use it. Farmers need to have the capacity to adopt and understand new technologies, and the system must develop to meet their needs and to enable them. Since most of the farmers in Africa are women, an important component of these systems will be including women in all parts of the proc-ess: education, capacity building, and technology innovation.

7

ENTREPRENEURSHIP

The creation of agricultural enterprises represents one of the most effective ways to stimulate rural development. This chapter will review the efficacy of the policy tools used to promote agricultural enterprises, with a particular focus on the positive, transformative role that can be played by the private sector. Inspired by such examples, this chapter will end by exploring ways in which African countries, subregional, and regional bodies can create incentives that stimulate entrepreneurship in the agricultural sector. The chapter will take into account new tools such as information and communication technologies and the extent to which they can be harnessed to promote entrepreneurship.

Agribusiness and Development

Economic change entails the transformation of knowledge into goods and services through business enterprises. In this respect, creating links between knowledge and business development is the most important challenge facing agricultural renewal in East African countries. The development of small and medium-sized enterprises (SMEs) has been an integral part of the development of all industrialized economies. This holds true in Africa. Building these enterprises requires the development of pools of capital for investment; local operational, repair, and maintenance expertise; and a regulatory environment that allows small businesses to flourish. The World Bank projects that agriculture and agribusiness will grow to be a US$1 trillion industry in Africa by 2030. Africa must review its incentive structures to promote these objectives and ensure that agriculture and agribusiness are at the top of the agendas for economic transformation.[1]

Across the continent, agriculture averages 24% of GDP; accounting for post-harvest activities, agriculture-related industry accounts for nearly half of all economic activity in sub-Saharan Africa.[2] Around the world, the prices of agricultural commodities have risen steadily since about 2000. In Africa specifically, urban markets are projected to quadruple in the next 20 years, creating an opportunity to commercialize and scale previously smallholder operations in order to feed urban populations. Today, sub-Saharan Africa spends $25 billion annually importing food into a region that holds about half the world's fertile and as yet unused land and uses only a tiny percentage of its renewable water resources.[3] Africa's food and beverage markets are projected to triple from their 2013 capitalization of $313 billion to about $1 trillion in just the next 20 years, according to World Bank projections. This growth will only be possible with adequate investment in agribusiness SMEs.

Small African firms engaged in agribusiness greatly outnumber the large players. According to a recent World Bank report on agribusiness in Africa: "In West Africa, 75% of agriculture-related firms are micro or small enterprises, 20% are semi-industrial, and 5% are industrial."[4] Value chains in many African countries are dualistic, featuring an informal chain serving lower-income consumers in domestic markets alongside a formal chain with more processing and stronger quality controls for higher-income, "'middle-class' domestic consumers or exports."[5] Larger businesses may dominate the high-end market, but in many sectors the vast majority of the volume moves through the smaller, less formal businesses. In Kenya, more than 95% of the fruit and vegetables produced move through their value chain with only smallholders and SMEs.[6]

A range of government policy structures is suitable for creating and sustaining enterprises—from taxation regimes and market-based instruments to consumption policies and changes in the national system of innovation. Policymakers also need to ensure that educational systems provide adequate technical training. They need to support agribusiness and technology incubators, export-processing zones, and production networks, as well as sharpen the associated skills through agribusiness education.

Banks and financial institutions also play key roles in fostering technological innovation and supporting investment in homegrown domestic businesses. Unfortunately, their record in promoting technological innovation in Africa has been poor. Capital markets have played a critical role in creating SMEs in other developed countries. Venture capitalists not only bring money to the table; they also help groom small and medium-sized start-ups into successful enterprises. Venture capital in Africa, however, barely exists outside South Africa and needs to be introduced and nurtured.

Much of the effort to promote venture capital in developing countries has been associated with public-sector initiatives whose overall impact is questionable.[7] One of the possible explanations for the high rate of failure is that many of these initiatives are not linked to larger strategies to create local innovation systems. Venture capital is only one enabling species in a complex innovation ecosystem.[8] It does not exist in an institutional or geographical vacuum and appears to obey the same evolutionary laws as other aspects of innovation systems.[9] It is therefore important to look at examples of geographical, technological, and market aspects of venture capital. The legal elements needed to create institutions are only a minor part of the challenge.

One critical starting point is "knowledge prospecting," which involves identifying existing technologies and using them to create new businesses. African countries have thus far been too isolated to benefit from the global stock of technical knowledge. They need to make a concerted effort to leverage expertise among their nationals residing in other countries. Such diasporas can serve as links to existing know-how, establish links to global markets, train local workers to perform new tasks, and organize the production process to produce and market more knowledge-intensive, higher value-added agricultural products.

Advances in communications technologies and the advent of lower-cost high-speed Internet will also reduce this isolation dramatically. The laying of new fiber-optic cables along the coasts of Africa and, potentially, the use of lower-latency satellite technology can significantly reduce the price of international connectivity and will enable African universities and research institutions to play new roles in rural development. The further development of Internet exchange points (ISPs) in East Africa, where they do not currently exist, is also important. ISPs enable Internet traffic to be exchanged locally, rather than transverse networks located outside the continent, improving the experience of users and lowering the cost to provide service.

Much is already known about how to support business development. The available policy tools include direct financing via matching grants, taxation policies, government or public procurement policies, advance purchase arrangements, and prizes to recognize creativity and innovation. These can be complemented by simple ways to promote rural innovation that involve low levels of funding, higher local commitments, and consistent long-term government policy.

For example, China's mission-oriented "Spark Program," created to popularize modern technology in rural areas, had spread to more than 90% of the country's counties by 2005. The program helped to improve the capability of young rural people by upgrading their technological skills, creating a nationwide network for distance learning, and encouraging rural enterprises to become internationally competitive. The program was sponsored by the Ministry of Science and Technology.[10]

There is growing evidence that the Chinese economic miracle is a consequence of the rural entrepreneurship that started in the 1980s. This contradicts classical interpretations that focus on state-led enterprises as well as receptiveness to foreign direct investment. The creation of millions of township and village enterprises (TVEs) in provinces such as Zhejiang, Anhui, and Hunan played a key role in stimulating rural industrialization.[11]

Over the past 60 years, China has experimented extensively with policies and programs to encourage the growth of rural enterprises that provide isolated agricultural areas with key producer inputs and access to post-harvest, value-added food processing. Despite a troubled early history, by 1995 China's TVEs had helped bring about a revolution in Chinese agriculture and had evolved to account for approximately 25% of China's GDP, 66% of all rural economic output, and more than 33% of China's total export earnings.[12]

Most of the TVEs have become private enterprises and focus on areas outside agricultural inputs or food processing. Agricultural support from TVEs remains relevant, however, particularly as a model by which other countries may be able to increase farmers' access to key inputs such as fertilizers and equipment, as well as value-added processing of raw agricultural products.

With few rural-urban connecting roads and weak distribution systems, the Chinese government moved to resolve these agricultural input and post-harvest processing constraints by creating new enterprises in rural areas. China's initial rural enterprise strategy therefore focused on the so-called five small industries that it deemed crucial to agricultural growth: chemical fertilizer, cement, energy, iron and steel, and farm machinery. With strong backward linkages between these rural enterprises and Chinese farmers, agricultural development in China grew substantially in the late 1970s and 1980s through farmland capital construction, chemical fertilization, and mechanization.

This expansion in agricultural productivity, coupled with high population growth, led to a surplus of labor and a scarcity of farmland. As a consequence, China's rural enterprises increasingly shifted from supplying agricultural producer inputs to labor-intensive consumer goods, for domestic and (after 1984 market reforms) international markets. From the mid-1980s to the 1990s, China's TVEs saw explosive growth in these areas while they continued to supply agricultural producers with access to key inputs, new technologies, and food-processing services. In 1993, 8.1% of total TVE economic output came from food processing, while chemicals (including fertilizer) accounted for 10%, building materials 12%, and equipment (including for farms) 18%.

The most successful TVEs were those with strong links to urban and peri-urban industries with which they could form joint ventures and share technical information; those in private ownership; and those with a willingness to shift from supplying producer inputs for farmers to manufacturing consumer goods for both domestic and international markets.

China's experience with rural enterprises confirms that they may provide a mechanism through which developing states can enhance rural access to key agricultural inputs such as fertilizers and mechanization, as well as value-added post-harvest food processing. Rural enterprises may make the most sense in areas where farm-to-market roads cannot be easily established to achieve similar backward and forward linkages. In addition to sparking agricultural productivity and growth, rural enterprises may also help provide employment for farm laborers displaced by agricultural mechanization. By keeping workers and economic activity in rural areas, China has helped expand rural markets, limit rural-urban migration, and create conditions under which it is easier for the government to provide key social services such as health care and education.

Despite the fact that TVEs enjoyed government support through financing and technical assistance, they also enjoyed a degree of autonomy in their operations. The emergence of rural markets in China not only contributed to prosperity in agricultural communities, but also provided the impetus for the modernization of the economy as a whole.[13] Furthermore, the TVEs also became a foundation for creating entrepreneurial leadership and building managerial and organizational capacity.[14]

In the absence of comprehensive interventions like China's Spark Program, some nonprofits and foundations are experimenting with promoting rural entrepreneurship by donating cows or other livestock to rural communities. Organizations like Heifer International provide cows, along with training for the recipients on how to raise them and profit from animal husbandry. The impact of these programs is relatively limited, however. For example, in Malawi, Heifer International is implementing a program alongside USAID designed to stimulate a dairy industry, but it serves only 180 smallholder farmers at present.

Such entrepreneurial initiatives will succeed in the absence of consistent and long-term policy guidance, on the one hand, and autonomy of action on the part of farmers and entrepreneurs, on the other hand. The latter is particularly critical because a large part of economic growth entails experimentation and learning. Neither of these can take place unless farmers and associated entrepreneurs have sufficient freedom to act. In other words, development must be viewed as an expression of human potentialities, not as a product of external interventions.

The Seed Industry

The seed industry in sub-Saharan Africa is informal in nature, with approximately 80% of farmers saving and replanting seeds from year to year. Yet better varieties—including high-yielding and hybrid crops—will increase productivity and income. To get these seeds into the hands of farmers, a better marketing and distribution system is needed. Emerging small and medium-sized seed enterprises have a comparative advantage in reaching this underserved market due to their size and market reach. These seed companies are the result of publicly supported agricultural biotechnology research and improved market opportunities for the private sector, and they are well positioned to promote food security and improve livelihoods among marginalized rural communities.

These companies, however, face financial and managerial capacity hurdles as well as complex bureaucratic and legal requirements. They offer unique benefits that could help grow the fledgling seed industry, but they need better access to credit and human resources to achieve their full potential.[15]

Maize is a staple in southern and eastern Africa, yet the amount of produce and the acreage of maize have not increased much over the years, even though the number of grain producers has quadrupled. However, the seed sector faces major challenges. Although less monopolized now, the seed sector in a majority of African countries is far from being efficient. The seed industry suffers from five levels of bottlenecks, producing an adverse effect on the maize seed value chain across the region. The first bottleneck is government political and technical policies. Import procedures, for instance, are cumbersome enough in Tanzania to dissuade seed import, while in Zimbabwe, during the economic crisis, the government banned seed exports.[16]

Second, establishing a seed company has a high initial cost, requiring access to credit; the company also needs qualified manpower. Third, the production of seed suffers from a lack of adequate and adapted input, from expensive production costs and lack of production credit, and from poor weather and unfavorable land policies. Fourth, poor infrastructure in the value chain, such as poor retail networks or sales points, jeopardize marketing and access to the farmers. Finally, farmers tend to have low demand for seeds.

Victoria Seeds is a good example of a Ugandan-grown seed supplier that has overcome the significant hurdles to the development of the seed sector to become Uganda's leading seed company. Founded in 2004 on the back of USAID guarantees, rather than loans from commercial banks, in less than 10 years Victoria Seeds grew to a turnover of $2.5 million, exporting to Rwanda, South Sudan, Tanzania, and the Democratic Republic of Congo.[17] Victoria Seeds has focused on improved seeds that deliver drought resistance and early yields. It sells primarily to small farmers, with a clientele that is overwhelmingly female.

Development of a Private Seed Industry in India

Millions of small-scale farmers in India live in harsh environments where rainfall is limited and irrigation and fertilizer are unavailable.[18] In these harsh areas, many farmers have long grown sorghum and pearl millet—hardy crops that can thrive in almost any soil and survive under relatively tough conditions. Production from these crops was low, however, and so were returns to farmers, until improved, higher-producing varieties were developed and distributed starting in the 1970s. Since then, a succession of more productive and disease-resistant varieties has raised farmers' yields and improved the livelihoods of about six million millet-growing households and three million sorghum-growing households. Although public funding was the key to developing this improved genetic material, it has been private seed companies that have helped ensure that these gains were spread to, and realized by, the maximum number of Indian farmers.

Three key interventions include increased investments in crop improvements during the 1970s; the development of efficient seed systems, with a gradual inclusion of the private sector in the 1980s; and the liberalization of the Indian seed industry in the late 1990s. By allowing farmers to grow the same amount of millet or sorghum using half as much land, these improved varieties have made it possible for farmers to shift farmland to valuable cash crops and thereby raise their incomes. The government played a key role in the last of these three innovations, the establishment of a private seed industry.

The first advances in millet and sorghum research in India resulted from the efforts of a range of government institutions. Joint research and testing efforts by state agricultural universities, research institutes, and experiment stations resulted in the release of a succession of pearl millet hybrids offering yield advantages. Since the mid-1960s, average grain yields have nearly doubled, even as much production of millet shifted to more marginal production environments. Production of pearl millet in India currently stands at nine million tons, and hybrids are grown in more than half of the total national pearl millet area of 10 million hectares.

At the beginning of the Green Revolution, the Indian government and key state governments decided that state extension services and emerging private seed companies could not distribute enough seed to allow for the large-scale adoption of new varieties. The government decided to create state seed corporations. "The Indian government, with the financial support of the World Bank and technical assistance from the Rockefeller Foundation, financed the development of state seed corporations (SSCs) in most major Indian states in the 1960s."[19] Gradually, these state seed corporations replaced state departments of seed production and formed the nascent foundations of a formal seed industry.

The institutional framework for the development of a seed industry emerged with the Indian Seed Act in 1966. The nascent Indian seed industry was heavily regulated under the act, however, with limited entry and formation of large private firms—domestic or foreign. Private seed imports for both commercial and research purposes were restricted or banned, ostensibly to protect smallholders from predatory corporate practices.

In 1971, India began deregulating the seed sector, relaxing restrictions on seed imports and private firms' entry into the seed market. This change, combined with a new seed policy in 1988, spurred enormous growth in private-sector seed supplies in India.

Sorghum and pearl millet breeding by private companies began around 1970, when four companies had their own sorghum and pearl millet breeding programs. By 1985 this number had grown to 10 companies. In 1981, a private company developed and released the first hybrid pearl millet. One major reason for the spurt in private sector growth was the strong public sector research on sorghum and millet. International agricultural research centers exchanged breeding material with public and private research institutions. National agricultural research centers and agricultural universities provided breeder seed not only to the national and state seed corporations but also to private seed companies to be multiplied and distributed through their company outlets, farmer cooperatives, and private dealers. For private firms, public institutions and state universities provided invaluable genetic materials, essentially free of charge.

Currently, the Indian market for agricultural seed is one of the biggest in the world. Today, more than 60 private seed companies supply improved pearl millet to small-scale farmers and account for 82% of the total seed supply, while more than 40 companies supply improved sorghum, accounting for 75% of supply. Many of these companies benefit not only from the availability of public research on improved pearl millet and sorghum but also from innovative partnerships that specifically aim to disseminate new materials to the private sector. The ultimate beneficiaries of this public-private system are the millions of small-scale farmers who grow sorghum and millet. Public research agencies contribute genetic materials and scientific expertise to improve crop varieties when the incentives for private-sector involvement are limited. Then, private companies take on the final development of new varieties and seed distribution—tasks to which they are often better suited than are public agencies. In this way, the benefits of crop improvements are delivered directly to farmers, who find them worthwhile enough to support financially.

All three elements of the Indian intervention to improve sorghum and pearl millet hybrids were important. First, the investments in public-sector plant-breeding and crop-management research were made by the national government, state governments, and international agricultural research centers. Second, the government invested in seed production in public and private institutions. The Indian government and state governments, with the help of donors, made major investments in government seed corporations that multiplied the seeds of not only wheat, rice, and maize, but also pearl millet and sorghum. New seed laws allowed small private-sector seed companies to enter the industry. Third, and most important, India liberalized the seed sector starting in the mid-1980s. The government opened the doors to investment by large Indian firms and allowed foreign direct investment in the sector. This change, coupled with continuing investments in public plant breeding and public-private partnerships, has continued to provide private firms with a steady stream of genetic materials for developing proprietary hybrids. The result is a vibrant and sustainable supply of seed of new cultivars that are drought tolerant and resistant to many pests and diseases.

Africa's Seeds of Development Program

Since 2003, the Seeds of Development Program (SODP) in eastern and southern Africa has aimed to improve access to affordable improved seed varieties for smallholder farmers. One major focus has been on management training for more than 30 small and medium-sized local seed companies, such as Victoria Seeds in Uganda, Freshco Seeds in Kenya, Kamano Seeds in Zambia, Qualitá in Mozambique, and Seed Tech in Malawi. SODP runs a fellowship program for management training with support from a grant from the UK Department for International Development. SODP is continuously expanding to new countries in the region to increase its network of fellows. Fellows receive generalized management training as well as specialized training geared toward the seed industry. Distance learning, travel grants to visit model seed companies, and a forum are all part of the program. In addition to its fellows program, SODP also created a research program. The research program conducts market research on the seed industry throughout the region.[20]

SODP management training is showing results. Companies chosen to participate in the fellows program sell seeds that are on average 20% cheaper than larger seed companies and reaching the intended client base—in fact, more than 80% of the sales go to smallholder farmers. Maize seed sales were also up by 54% from 2006 to 2007; employment increased by 19%; and revenue increased by 35%. SODP companies typically offer a wide range of seeds that include high-yield, disease- and/or drought-resistant, and herbicide-tolerant varieties.

The SODP network facilitates partnerships and sharing of business practices after the fellowship ends. SODP's link with the Alliance for a Green Revolution has also proven helpful for the small and medium-sized seed companies. SODP's follow-up program will focus on connecting seed traders and full-service seed companies to allow each to exploit their market niche.[21]

Food Processing

Transformations in the food-processing sectors of developing countries are increasingly seen as strategic from the point of view of export earnings, domestic industry restructuring, and citizens' nutrition and food security.[22] The widespread adoption by developing countries of export-led growth strategies has drawn attention to the economic potential of their food-processing sectors, particularly in light of the difficulties faced by many traditional primary commodity export markets. Food processing can be understood as post-harvest activities that add value to the agricultural product prior to marketing. In addition to the primary processing of food ingredients, it includes, therefore, final food production as well as the preparation and packaging of fresh products. To better understand the role of food processing in African agricultural development, this chapter will examine several cases of successful African food-processing start-ups, as well as the role that new technologies (particularly radio and video) can play in teaching farmers how to add value with post-harvest processing.

Homegrown Company, Ltd., Kenya

Homegrown Company, Ltd., which was founded in 1982, processes and exports packaged horticulture produce from Kenya, primarily to the United Kingdom. The goal was to package the produce at the source in order to avoid repackaging abroad. In addition to sourcing its own produce, Homegrown partnered with local farmers to boost its exports. Contract farmers account for approximately 25% of the produce.

To ensure an adequate supply, Homegrown enters into a contract with these farmers that specifies the type, quality, and quantity of produce that the farmer is required to supply and the price that they will receive. This allows farmers to plan their schedules accordingly and to purchase inputs on credit in order to meet Homegrown's requirements. The company now has a network of reliable producers in addition to its own production units that allow it to maintain high production levels and its customer base.

Smallholder farmers benefit from these contracts as well. Risk is decreased significantly because farmers are guaranteed both a market and a predetermined price for their produce, depending on prearranged quality indicators, and they can better plan their seasons with greater foresight. Furthermore, farmers benefit from new technologies and farming techniques, and have access to inputs on credit from Home-grown that they might not otherwise be able to use.[23]

Blue Skies Agro-Processing Company, Ghana

Blue Skies is another successful agro-processing company located about 25 kilometers from Accra. Blue Skies processes fresh fruits destined for supermarkets in some European markets. Fruits include pineapple, mangoes, watermelon, passion fruit, and pawpaw. Most fruit is produced and processed in Ghana, but the company fills gaps by importing fruit from South Africa, Egypt, Kenya, Brazil, and the United Kingdom. The company started with 38 workers and has increased the workforce to 450; 60% are permanent employees. The produce is processed so that it meets the standards of the European Retailer Partnership Good Agricultural Practices (EUREGAP); in fact, the company only sources fruit from farmers who are EUREGAP certified—and the company helps in the certification process. The company provides training and extension services to farmers, in exchange for higher-quality fruit that meets the standards. Blue Skies offers inputs and equipment via interest-free loans to farmers. The company also typically offers higher prices and pays farmers promptly, thereby encouraging farmers to produce higher-quality fruit. As such, Blue Skies can now process 35 tons per week—a huge improvement over the one ton per week of which it was initially capable.[24]

Blue Skies also provides its dedicated farmers with credit and has worked to improve road infrastructure near farms and to enhance access by company trucks.

Cassava Bread in Nigeria[25]

Since its debut in the late 1600s on Portuguese trade ships from Brazil into Nigeria, cassava has become an important crop that accounts for 30%–50% of all calories consumed in southern and central Nigeria.[26] It is produced predominantly by small farmers with 1–5 hectares of land, intercropped with yams, maize, or legumes in the rainforest and savannah areas of southern, central, and recently northern Nigeria, and processed locally by rural women. Nigeria is now the largest producer of cassava in the world, with a total production of 55 million metric tonnes of fresh cassava roots in 2014, but it is not even among the top 20 countries exporting processed cassava; Thailand is currently the largest exporter.[27] Reasons include Nigeria's cassava sector having low productivity, leading to high costs per unit production, and an inability of Nigerian cassava products to compete with imported substitutes, resulting in a lack of demand for cassava by industrial users who prefer to import raw materials.[28]

In 2003, the second wave of cassava innovation began in Nigeria. Driven by President Obesanjo, the Presidential Initiative on Cassava sought to make cassava a commodity crop and foreign exchange earner by transforming its role in the economy beyond that of a "traditional" food crop and into a staple that could compete with wheat, maize, and rice. In addition to increasing the total amount of cassava being produced in the country, this initiative focused on improving cassava-based food products. There was now an institutional incentive to use the research from the 1980s to bring a product to market. Researchers using the composite flour had shown that mechanical leavening rather than bulk fermentation for the ripening of the dough and a blend of 60% wheat flour, 30% cassava starch, and 10% soybean flour produced a bread of good quality, almost equal to the incumbent wheat-flour bread in volume, appearance, and taste.[29] Joint research had finally succeeded in using high-quality cassava flour (HQCF) as a viable partial substitute for wheat flour in baking bread. However, the acceptability of cassava as a substitute by consumers was still problematic; cassava-based bread was still considered inferior to wheat-based bread and the use of cassava in a composite flour needed improvement.[30] When President Yar'Adua succeeded Obesanjo, cassava was once

again off the government's agenda, but joint research con-
tinued to take place between the International Institute of
Tropical Agriculture (IITA) and Nigeria's Federal Institute of
Industrial Research

By the time President Goodluck Jonathan officially took
office in 2011, a 20% HQCF loaf had been developed that met
standards of color, crust, taste, texture, and aroma. However,
there had been very little commercial prospects for the technology due to a variety of cultural and institutional barriers.[31] These included the fact that millers were only willing to
mix composite flour containing 5% cassava flour and because
bakers claimed that the loaves made from any higher percentage of cassava flour did not meet customers' quality standards.[32] This complaint was largely due to both flours being
of insufficient quality and a lack of know-how, rather than a
problem with the technology itself. The emphasis of the innovation system therefore had to move from the "invention"
stage to focus on the "production" and "sustained use" stages
of the innovation system.

In 2011, the Nigerian federal government initiated the Cassava Transformation Agenda. A fundamental aspect of the
agenda was an import substitution policy to halt Nigeria's reliance on imported wheat: the Cassava Bread Development
Policy. In May 2012 this bill, making it compulsory for all bread
to contain 10%–40% cassava flour, was brought to the House of
Representatives, but it was met with opposition. The Minister
of Agriculture said that it would save Nigeria approximately
$252 million in wheat imports while building a domestic in-
dustry, but opposition came from the wheat importers lobby,
as well as the Association of Master Bakers, Confectioners and
Caterers, who said that they did not have the means or technology to make the bread. Added to this were controversies
over cassava from a health perspective, with its anti-nutritional
properties being touted together with an argument that it was
unsafe for diabetics, of which Nigeria has a high percentage in
the population.

Although the bill did not pass, in July 2012 the Federal Executive Council passed fiscal policies aimed at promoting the adoption of cassava bread. These include a 65% levy on the importation of wheat flour, 15% on wheat grain that would be paid with the 35% duty on the commodity. Cassava flour imports were banned and there was duty-free import of equipment for processing HQCF. Bakeries with 40% HQCF inclusion were also to receive a 12% corporate tax rebate. Part of the funds generated from this initiative would go toward a Cassava Bread Development Fund to build capacity in the agricultural sector through expertise in producing, processing, and exporting cassava. Along with this, the tariff on an enzyme critical to the processes of making cassava bread would be removed.

As of December 2014, a composite flour policy to create incentives for wheat millers and to other actors in the cassava bread value chain has been drafted. The policy calls for the mandatory inclusion of 10% cassava or sorghum flour in all bread produced in Nigeria by December 2016; the current universal rate of inclusion of HQCF in cassava is 2% due to limited quantities of HQCF. In exchange, a reduction of 5% in the import levy on wheat grains is provided to these millers. The policy has started to yield results; two of the largest millers in the country, Flour Mills of Nigeria and Honeywell Mill, which together account for 70% of all flour milled in the country, have launched 10% cassava flour brands, making composite flour more easily accessible to bakers, although bakers still have to buy additional HQCF to reach the desired 20% inclusion. The effort to reduce wheat imports is also paying off: wheat imports declined from an all-time high of 4,051,000 tonnes in 2010 to 3,700,000 tonnes in 2012, and continue to decrease as HQCF production is accelerated.

Cassava bread has been available in stores since October 2012 when Park and Shop, one of the largest supermarkets in the country, launched its own 25% cassava flour bread at its outlets in across the country. Other corporate bakers, including Sweet Sensation, Grand Square Bakery, Exclusive Stores Bakery, ShopRite and Imperial Bakeries, quickly followed with their own 20% cassava bread product. In addition to the six large industrial bakers, 23 master bakers (small-size bakeries) also successfully launched 20% cassava bread across the country. The HQCF technology is also not limited to cassava bread; indeed, it is almost easier to experiment with other baked goods because customers do not require them to conform to cultural standards in quite the same way as they do bread, which is a staple food. Other confectionaries, such as jam doughnuts, meat-pies, and sausage rolls, as well as croissants and cake, have also been developed from the composite flour and some are available on the market.[33] The private sector has recognized potential in the market for other baked goods made from HQCF, and there has been a much quicker increase in the uptake of these baked goods than there has been of cassava bread, even though there has been no direct government push to market these products.[34]

Orphan crops such as cassava often suffer from being designated as inferior or "backward"; cassava is still viewed as "the famine crop," whereas major crops like wheat and rice are seen as progressive or something to aspire to being able to buy.[35] Innovations that can break these stereotypes and create "niche" markets for these products must form a critical aspect of any orphan crop innovation agenda. However, the bridging of formal innovation systems with the innovative potential embedded in the traditional knowledge sources that are often associated with orphan crops will require ethical institutions that recognize the power dynamics and inequities that could accompany interventions. This was distinctly lacking in the cassava example because it was a top-down initiative, driven by the perceived importance of developing Nigeria's cassava value chain. Although still a nascent field, the importance of developing innovation systems for orphan crops is going to become increasingly important as we face growing challenges in the food system to meet food security requirements.

Learning from existing cases is the best way to move forward sensibly and to channel resources in the manner that will see sustainable and equitable outcomes, rather than the mere delivery of yet another food product.

Communication Technology in Food Processing

Communication technology is another way of encouraging agricultural innovation and transformation in the sector. Media, radio, and video are all appropriate methods to reach farmers and agribusinesses alike. Videos are particularly useful when it comes to food processing. In Benin, women are more likely to apply innovative food-processing techniques they learned from videos than from traditional training workshops. This method of reaching farmers is not capital-intensive, and could fill the much-needed gap in extension services as well. Video is also a more efficient way to reach farmers than traditional training workshops.

Video provides "a powerful, low-cost medium for farmer-to-farmer extension and for exposing rural communities to new ideas and practices."[36] A recent study "examined the impacts of educational videos featuring early adopting farmers demonstrating the use of new technologies and techniques. The study found that when women watched videos featuring fellow farmers demonstrating new techniques, they showed better learning and understanding of the technology and creatively applied its central ideas. Innovation levels of 72% were recorded in villages where videos were used to introduce women to improved rice-processing techniques."[37] This can be compared to 19% innovation among farmers who attended training workshops. When women who had attended training workshops watched the videos, the innovations increased to 92%. The videos stimulated creativity among rural farmers.

Drawing on lessons from a similar initiative in Bangladesh, the Africa Rice Center is using local language videos to train farmers on various aspects of rice production and processing in

Benin, Ethiopia, Gambia, Ghana, Nigeria, and Senegal, among other locations.[38]

By 2009, rice videos had been translated into 30 African languages, and had been adopted by more than 400 community-based organizations and viewed by 130,000 farmers to strengthen their own capacity in rice technologies. The videos are disseminated through mobile cinema vans or local organizations, reaching three times as many farmers as in-person training workshops. Partner organizations in various countries are combining the videos with radio programming to reinforce the lessons and knowledge.

One Guinean radio station, Radio Guinée Maritime, has aired interviews with farmers involved in this program, reaching some 800,000 listeners; Gambia, Nigeria, and Uganda have similar initiatives. To effectively capitalize on the potential of this technology in Africa, proponents advise broadening the dissemination of innovations beyond those developed by the traditional research and extension systems to include localized farmer innovations as well.[39]

Mobile phone apps represent another cost-effective and efficient way for smallholder farmers to access up-to-date information.[40] Kenya first developed an information portal called Infonet that aimed to provide farmers and extension agents with data from Kenya's agricultural research ecosystem. However, a majority of farmers lacked Internet access. In 2010, an innovative platform called "iCow" sought to use increasingly popular and accessible mobile technology to bridge the gap. iCow started out as "Mkulima F.I.S.H," or Farmer Information Service Helpline, as a way to educate farmers over the phone. Mkulima F.I.S.H. was designed to be an off-grid agricultural wiki for mobile phone users. One component of the platform was the cow calendar, which eventually won an Apps4Africa competition. Subsequently, a team of developers and a customer service team worked closely with farmers on the ground so that farmer challenges were prioritized, and smart solutions were designed and developed to address these problems. Today, iCow offers a variety of products. The original cow calendar extends to poultry calendars, and subscription products enable farmers to sign up for targeted information on livestock or crops. In addition, a pay-as-you-go service enables farmers to access relevant information as and when they require it. iCow also gives farmers access to other relevant stakeholders in their ecosystem, from credit providers to input providers to buyers. As of November 2014, iCow has more than 160,000 users and a database of 500,000 farmers.

Data and user surveys show that iCow is improving farmer productivity by reducing farming risks. Dairy farmers are increasing milk yields after as little as 3 months on the subscription product, where they learn best dairy practices. The increases in milk are up by 2–3 liters per animal within the first 3 months, and over time this is seen to increase to 4–6 liters per animal. Using iCow accounts for 30% to 150% of improved yields. Data from iCow poultry farmers is also showing reduced chick and bird mortality, increased egg production, as well as increases in batches of birds over time. Interestingly, farmers who have never reared poultry are now beginning to do so. In addition to increased yields, less tangible yet equally significant changes are happening, including better animal health, which is a precursor to improved yields.

iCow continues to innovate and produce tools that reduce farming risk, such as "Best Mbegu for You" (Best Seed for You), where farmers can find seeds that are relevant for their geographical location, and "Know Your County Soils," where farmers can learn of the deficiencies within the soils in their general area before being pointed to soil-testing services where they can get cost-effective access to testing and soil analysis.

Social Entrepreneurship and Local Innovations

Social enterprises are emerging as major economic players worldwide.[41] Their role in African agriculture is growing.

One Acre Fund

An example of such an initiative is One Acre Fund, a nonprofit organization based in Bungoma (western Kenya) that provides farmers with the tools they need to improve their harvests and feed their families.[42] Life-changing agricultural technologies already exist in the world; One Acre Fund's primary focus is on how to distribute these technologies directly to smallholder farmers, adapted to fit smallholder needs, and ensuring wide-spread adoption. One Acre Fund currently serves 200,000 farm families (with 700,000 children in those families) in Kenya, Rwanda, Burundi, and Tanzania.

From the beginning, One Acre Fund talked to farmers to understand what they needed to succeed. Farmers reported lack of access to credit, quality farm inputs, markets, and information on effective farming technique as significant barriers to achieving prosperity. One Acre Fund's model removes these barriers by addressing the full agricultural value chain. When a farmer enrolls with One Acre Fund, she joins as part of a group of approximately 12 farmers. She receives an in-kind loan of seed and fertilizer, which is guaranteed by her group members. One Acre Fund delivers this seed and fertilizer to a market point within two kilometers of where she lives. After the delivery of inputs, staffers known as "field officers" provide farmers' groups with on-site trainings on land preparation, planting, fertilizer application, weeding, composting, and other farming best practices. These trainings are standardized across One Acre Fund's entire operation and include interactive exercises, simple instructions, and group modeling of agriculture techniques. For instance, after a field officer teaches a group of farmers how to use a planting string to space rows of crops, he or she asks them to model the technique in the field so that he can offer immediate feedback.

Over the course of the season, the field officer monitors the farmer's fields. At the end of the season, he or she teaches the farmer techniques for effectively harvesting and storing her crop, and offers trainings on how to maximize profits by carefully timing the sale of her surplus. While farmers have the option of flexibly scheduling loan repayments incrementally over the course of the season, final loan repayment is several weeks after harvest. On average, 98% of One Acre Fund farmers repay their loans in full and on time.

Before joining One Acre Fund, many farmers in Kenya were harvesting five bags of maize from half an acre of land. After joining One Acre Fund, their harvests typically increase to 12 to 15 bags of maize from the same half acre of land. This represents a doubling in farm profit per planted acre—twice as much income from the same amount of land.

The field officer is the most important part of the One Acre Fund operating model. Field officers often come from the same communities in which they work. One Acre Fund consciously chooses not to hire university-educated horticulturists for its field staff (like most NGOs) because most of the information farmers need to know is encapsulated in a few simple lessons. They instead choose to hire local people—many of whom are also farmers—who demonstrate leadership potential and a deep desire to serve their own communities.

One Acre Fund's field officers each work with 150–220 farmers (depending on geography), and they visit each of their farmers on a weekly or biweekly basis. At these meetings, they conduct trainings, check germination rates, troubleshoot problems in the field, and collect loan repayments. Over the course of the season, One Acre Fund's field officers cultivate a strong bond with the farmers they serve. They actively solicit feedback from farmers about farming techniques and One Acre Fund's program. In turn, One Acre Fund farmers have a deep appreciation for how knowledgeable their field officers are and how hard they work to serve their customers. Many farmers call their field officers "teacher."

Tomato Jos

Another promising local innovation is Tomato Jos in Nigeria.[43] Tomato Jos is an agricultural production company that believes in the power of making local food products for local consumption. It is a for-profit social enterprise that produces tomato paste in Nigeria for the domestic market, and it sources raw materials from smallholder farmers.

Though tomatoes and tomato paste are staples in Nigerian cooking, it seemed that the thousands of subsistence tomato farmers were unable to sell a significant amount of their crops—whole fields of tomatoes were rotting on the vine for lack of a market. Each year, rural farmers grow $2 billion worth of fresh tomatoes, but 50% rot before they can reach the market due to poor infrastructure and the delicate nature of tomatoes. As a result, Nigeria imports almost $500 million of bulk-packaged tomato paste each year (retail value of $1.5 billion), but smallholder farmers are cut off from this large and growing market. Tomato Jos saw an opportunity to provide farmers with a steady income, reduce agricultural waste, and decrease Nigeria's dependence on imported food.

A well-located, commercial tomato-processing operation focused on continuous production can increase smallholder farmer incomes by five. Tomato Jos's model links farmers directly to paste production, simultaneously reducing poverty for a vulnerable subset of the population and decreasing the need to import a dietary staple in the local cuisine. At scale, Tomato Jos will operate three business lines: (1) farm and agricultural center with farmer education and bundled inputs to help smallholder farmers grow and harvest crops more efficiently; (2) logistics and supply chain support to navigate the "last mile" to farms and safely bring produce to the factory; and (3) food-processing and packaging facility that prepares branded paste for Nigerian markets.

Retail packaged tomato paste is an important market in Nigeria. There is currently no tomato paste processing capacity in the country. Ninety percent of imported paste comes from China and is a low-quality product. Existing brands target consumers without segmenting the market, making little to no effort to differentiate themselves. Tomato Jos's vertically integrated value chain offers the most advanced inputs to farmers with support and training to maximize their full potential, and access to a guaranteed market. It can then offer end consumers a higher quality locally branded product at competitive prices.

Success is measured in terms of financial returns, job creation, and direct farmer impact. Smallholder farmers will see yields increase through advanced farming practices, high quality seeds and inputs, and individualized hands-on support. They will also experience less post-harvest loss. Estimates suggest that 40%–60% of tomatoes are "lost" after harvest. By improving post-harvest transport and providing a steady market, Tomato Jos will substantially decrease the percentage of tomatoes lost after harvest. At scale, the company will work with 1,000 farmers, increasing their collective income by $4 million.

Aldeia Nova

Aldeia Nova is an agro-industrial center portfolio company investment of Vital Capital Fund, one of the world's largest private equity impact investment funds. The company follows an agro-communal model based on the Israeli "Moshav" concept, a community-based agricultural structure that helped Israel build a modern agricultural sector, maximizing its human and natural resources.[44]

Aldeia Nova fuses agricultural production with service provision and social development by establishing and operating a large-scale, agro-industrial center in the Waku-Kungo region of Angola. The center supports several communities of Angolan farmers and their families by providing essential production support—animal feed, mechanical equipment, processing and packaging facilities, professional services, training, and infrastructure. It also provides 100% off-take and distribution for the resulting production, which is then sold downstream by the center.

Aldeia Nova supports the development of modern, skilled, self-sustaining communities, bringing new life to this developing region of Angola and, in turn, the whole country. Vital Capital's investment agreement includes an exclusive management contract to operate Aldeia Nova.

As in many other African countries, food security in Angola is a major issue, especially in rural areas. The region where Aldeia Nova is located—Waku Kungo—is very fertile and provided a large part of the country's food supply during colonial times as well as generating exports. Both the need and the opportunity existed to re-create the region's once vibrant and successful agricultural production.

Aided by the consistently growing demand for locally produced foodstuffs such as eggs, poultry, dairy products, and crops, the villages supported by Aldeia Nova are experiencing rapid growth in production volumes and prices, and the agro-industrial center that serves as the heart of the system is playing a key part in that improvement.

Today, an estimated 7 million Angolans, or 40% of the population, are undernourished. Aldeia Nova tackles this challenge head-on by providing a major income source for local populations and increased food security through the establishment of local agriculture brands, proudly labeled as "made in Angola." Aldeia Nova's social impact also builds on the creation of economic and commercial literacy among the local farmers and the broader community, which enables a rapidly developing economic ecosystem.

The result is an economic engine and regional powerhouse of agricultural production. Presently, Aldeia Nova produces more than 250,000 eggs per day, representing more than 40% of the national production in Angola. Annually, 730,000 liters of milk, 1,700 tons of soya, and 1,800 tons of maize are produced, and local farmers' income has increased by a factor of 10 since Aldeia Nova's central agro-industrial processing center opened. More than 1.5 million customers are estimated to have purchased Aldeia Nova-branded products. Over 600 local

210 THE NEW HARVEST

farmers are engaged in the Out Growers scheme, under clear contract. Above 80% of the company's purchases are local (over $15 million of local purchases), and above $11 million are purchases from small suppliers, including the farmers. In addition, Aldeia Nova hires more than 360 direct employees who are provided training in different areas. Accounting for the scale and depth of impact, Aldeia Nova plays a great role in the revival of the area, and its impact spans throughout the country.

By virtue of the economic and social impact of Aldeia Nova, the company has received a platinum rating from the Global Impact Investing Rating System, the most recognized third-party impact assessment.

Local Innovations

Local innovations represent one of Africa's least recognized assets. For more than 20 years, India's Honey Bee Network and Society for Research and Initiatives for Sustainable Technologies and Institutions have been scouting for innovations developed by artisans, children, farmers, women, and other community actors. They have built a database of more than 10,000 innovations.[45]

To further the work, India's Department of Science and Technology created the National Innovation Foundation (NIF) in 2000. Its aim is "providing institutional support in scouting, spawning, sustaining and scaling up grassroots green innovations and helping their transition to self supporting activities." NIF has so far filed over 250 patent applications for the ideas in India, of which 35 have been granted. Another seven applications have been filed in the United States, of which four have been granted.

To facilitate the commercialization and wider application of the innovations, NIF works with institutions such as the Grassroots Innovations Augmentation Network, which serves as a business incubator. Some of the objectives of the Honey Bee Network later became part of the work of the Indian Prime Minister's National Innovation Council. Africa's diversity in agricultural and ecological practices offers unique opportunities for creative responses to local challenges. Such responses form a foundation upon which to supplement formal institutions with entrepreneurial activities driven by local innovations.

Crop Diversification

As nations and development agencies work to promote agribusiness entrepreneurship, they must avoid the impulse toward monocultures that focus exclusively on cash crops or high-yielding grains. Agriculture-dependent economies must mitigate their downside risks of swings in the commodities market. Diversification of agribusiness can also bolster food security and nutrition.

One promising target for diversification from a food security perspective is breadfruit.[46] Breadfruit (*Artocarpus altilis*) is a long-lived, perennial tree that is well adapted to a wide range of tropical environments. It is unique in producing a starchy fruit equivalent to annual staple field crops such as rice, maize, cassava, and sweet potatoes. Breadfruit has been grown in Oceania for more than 3,000 years, where it is planted in mixed gardens, around homes, and in villages. On some islands, it serves as an integral part of complex, multispecies agro-forests, with entire hillsides managed around the trees. These food forests are a model for sustainable food production systems in the tropics, requiring minimal inputs of labor or materials.

The starchy fruit is high in energy and carbohydrates and is a good source of fiber, minerals, and vitamins. A 1,000 calorie serving can provide over 100% of carbohydrate and fiber, over 50% of potassium and magnesium, over 20% of protein, Vitamin C, iron, calcium, phosphorus, and over 8% of Vitamin B9 (folic acid) of the daily recommended dietary allowances (RDA). Some varieties are also a good source of pro-vitamin A carotenoids.[47]

This remarkably productive tree produces up to 450 pounds of fruit each year.[48] "The fruit packs 121 calories in a half-cup serving and is rich in fiber, potassium, phosphorous, calcium, copper and other nutrients. Its texture and yeasty odor remind some people of fresh bread."[49] Breadfruit is a traditional staple food in the Pacific Islands, and from there spread to Africa and the Caribbean recently. It is remarkably tolerant to a range of agro-ecological conditions, producing fruit at a relatively wide range of altitudes, rainfalls, and temperatures.

Breadfruit trees can begin bearing in two and a half to three years and are productive for many decades. A tree can readily bear 250 fruit (each averaging 1.2 kg), or more, annually. A one-hectare planting of 50 trees could produce up to 5.6 tonnes of fruit after seven years, approximately 1.7 tonnes/hectare of dry matter.[50] This compares favorably to average global yields for corn (4 tonnes), rice (4.1 tonnes), or wheat (2.6 tonnes).

The major limitations on greater utilization of breadfruit are related to the perishability of the fruit, the seasonal nature of the crop, and the availability of good-quality planting material. Breadfruit must be consumed soon after harvest, or processed into more shelf-stable forms such as flour, meal, fermented dough, beverages, and other value-added products.[51]

Only a fraction of breadfruit's diversity is available in other tropical areas. In the late 1700s, a few seedless Polynesian varieties, and related species, seeded Artocarpus camansi (breadnut) were introduced to the Caribbean, and spread from there to Central and South America, Africa, Asia, and the Indian Ocean islands. Introduction and dissemination of new breadfruit varieties from the Pacific has not occurred until recently.

More than 120 varieties from throughout the Pacific are conserved and studied in the world's largest breadfruit repository at the National Tropical Botanical Garden's Breadfruit Institute in Hawaii. Extremely nutritious, productive varieties that can provide an extended fruiting season have been identified and selected for global distribution. The traditional method of propagating breadfruit by root suckers is labor intensive, inefficient, and slow. The Breadfruit Institute, researchers at the University of British Columbia Okanagan, and partners in the private horticultural sector have successfully developed new micropropagation methods to produce and distribute healthy and vigorous plants.

There is a renewed interest in breadfruit for food security, sustainable agriculture, agroforestry, reforestation, and income generation. Many African countries are particularly interested in introducing new varieties to complement the one or two varieties that are currently grown. Since 2009, more than 50,000 new breadfruit trees have been planted in 31 countries. Pilot projects have been established in Ghana, Kenya, Liberia, Tanzania, Nigeria, Madagascar, Mozambique, Rwanda, and Zambia. Expanding breadfruit tree plantings will require support and investment from government agencies and an integrated strategy utilizing the strengths and expertise of local research universities, farmer-based organizations, public health professionals, and the entrepreneurs involved in food production, value-added products, and marketing.

Scaling Up

One of the most pressing challenges in agribusiness in Africa is scaling up start-ups. Much of the policy emphasis today is on promoting start-ups, with little attention going to scaling up. There is an implicit assumption among policymakers that creating the right policy environment for entrepreneurship will not only foster business incubation, but also automatically enable the more successful enterprises to grow. The reality is that scaling up is a nonlinear process that requires different sets of interventions and measures. In industry, for example, scaling up can be achieved through mergers and acquisitions, with greater possibility for the relocation of productive assets. In agriculture this is not usually the case because the most critical productive asset, land, is fixed in place.

In addition to these challenges, scaling up in agribusiness needs to take into account the structure of African agriculture, which is dominated by small-scale farmers as well as small-scale processors and distributors. Such conditions require alternative business models that balance between the autonomy of decentralized production and the need to scale up for purposes of marketing and distribution. One way to pursue this objective has been to adapt the franchise model to African agriculture.

Babban Gona Agricultural Francise

One example that illustrates the use of this approach is the Babban Gona agricultural franchise in Nigeria.[52] Across the globe, successful farmer organizations have played a critical role in enabling farmers to increase their productivity and profitability. However, in Nigeria, Africa's largest economy, despite agriculture contributing 22% to GDP and 70% to the workforce, there is a lack of strong farmer organizations that provide quality income-enhancing services to members. To address this vacuum, in 2012 Doreo Partners, a leading impact investing firm in Nigeria, launched Babban Gona ("Great Farm" in Hausa), an innovative agricultural franchise model that unlocks the power of grassroots-level leaders to operate strong farmers' organizations.

The Babban Gona agricultural franchise model is a partnership between Babban Gona (the franchisor) and a grassroots farmer organization comprising 3–10 smallholder farmers, called a "Trust-Group" (the franchisee). A Trust-Group is a model for operating a farmer organization, where members receive a standard set of products and services to optimize net farm incomes and yields.

Babban Gona has scaled 100-fold in four years, growing from 16 Trust-Groups in 2012 to 1,600 Trust-Groups for the 2015 season, with a current demand of 1,000 Trust-Groups per week applying to join Babban Gona. This growth has been accelerated due to the fourfold increase in member net incomes, above the income of the average Nigerian smallholder farmer. This increase in net income has been driven by access to knowledge and quality agricultural inputs and services, leading to increased yields that are up to 5.5 times the national averages and an Enhanced Warehouse Receipts Program that has increased maize prices attained by members by up to 37%.

The three key success factors for a farmer organization are (1) committed leadership, (2) professional management, and (3) investment to scale. Hence, Babban Gona begins by identifying smallholder farmers with the requisite skills to be leaders of farmer organizations, leveraging an innovative psychometric testing platform. Once these leaders are identified and their organization formed, Babban Gona provides the requisite professional management through a holistic set of products and services that effectively de-risks the members of these franchise farmer groups, simultaneously unlocking the required investment for members to scale their farming operations.

The key products and services provided by Babban Gona are training, credit, agricultural inputs, and marketing services. Babban Gona begins by providing Trust-Groups with training in leadership, business skills, and agronomy. This training is provided through a combination of classroom programs offered via its Farm University program and field training provided by its field officers.

To attract investors to provide capital for Babban Gona to on-lend to members, Babban Gona utilizes its proven risk-mitigation mechanisms, which include the development of Nigeria's first weather index insurance program, all of which have enabled a loan repayment rate of 99.8%. Such high repayment rates enable Babban Gona to raise cost-effective capital and deliver it sustainably at half the cost of traditional microfinance.

The credit provided by Babban Gona comes in-kind, through the provision of a holistic set of high-quality agricultural inputs and services. The products and services are delivered via a highly efficient distribution network at competitive prices and include soil analysis, seeds and other inputs, land preparation services and harvest threshing services, and even needles, thread, and bags for packaging the products.

The Babban Gona franchise model successfully integrates the strengths and overcomes the weaknesses of traditional farmer organization and out-grower models by providing the professional management and investment to scale found in out-grower models, while maintaining the high levels of trust found in farmer cooperatives. This high level of trust is driven by the fact that Babban Gona does not buy produce from farmers but instead markets the produce on their behalf.

Looking to the future, the Babban Gona model is designed to attract a new generation of farmers into agriculture by specifically addressing the unique challenges affecting young farmers. Young smallholder farmers are at a particular disadvantage compared to older farmers, in terms of potential net incomes generated from their farming operations. First, younger farmers tend to have smaller landholdings, and thus limited scope for income generation, particularly due to low yields. Second, younger farmers, with young families, have limited access to the subsidized labor that older farmers often have, with teenage children who can support them in their farm work. Third, young farmers usually have smaller savings and fewer assets, leaving them in no position to adequately finance the procurement of critical yield-enhancing technologies. The model addresses each of these by increasing yields, financing access to labor-saving products and services, and unlocking critical working capital. For this reason, compared to the national average age of farmers of 50–55 years old, Babban Gona has a youth membership of 47%. With the goal of rapidly and permanently increasing the productivity and incomes of one million smallholder members by 2025 to improve their livelihoods, Babban Gona's model will play a critical role in addressing Nigeria's spiraling youth unemployment, with an estimated 60% of youth unemployed and 80 million new youth entering an oversaturated job market in the next 20 years.

Conclusion

Despite strong growth in the private seed sector in Africa over the last decade, most of Africa's millions of small-scale farmers lack easy access to affordable, high-quality seeds. Seed policies and regulations currently differ across African countries, limiting opportunities for trade and collaboration. However, efforts are under way to develop regional trading blocs in the seed industry. For example, across the 14 Southern Africa Development Community countries, seed industry stakeholders have been formulating a single policy document to enable companies to move seed and breeding material across national borders, register varieties more easily, and market their products regionally. A parallel initiative is under way for East African Community countries. These efforts need to be finalized in eastern and southern Africa, replicated across all Africa's subregional organizations, and complemented by parallel efforts in the African Union.

Like the formation of African seed companies, the creation and spread of value-added food-processing enterprises could help African farmers retain a higher portion of the profits from the materials they produce. Food processing could also help reduce the threat of hunger by increasing the number of protein- and vitamin-rich products provided by the local market, as well as improve local incomes by tapping into international markets to get much needed export revenues from agriculture. Unlike the situation with seeds, growth in food processing will require fewer changes in government and regional policies. The key change will need to come in the areas of capital, so that it is easier for individuals and companies to invest in the infrastructure, equipment, and training necessary to enter the food processing industry.

8

GOVERNING INNOVATION

African countries are increasingly focusing on promoting regional economic integration as a way to stimulate economic growth and expand local markets. Considerable progress has been made in expanding regional trade through regional bodies such as the Common Market for Eastern and Southern Africa and the East African Community. There are six other such Regional Economic Communities that are recognized by the African Union as building blocks for pan-African economic integration. Thus far, regional cooperation in agriculture is in its infancy and major challenges lie ahead. This chapter explores the prospects of using regional bodies as agents of agricultural innovation through measures such as regional specialization. The chapter will examine ways to strengthen the role of the RECs in promoting innovation. It adopts the view that effective regional integration is a learning process that involves continuous institutional adaptation.[1]

Through extensive examples of initiatives at the national or cross-border levels, this chapter provides cases for regional collaboration or scaling up national programs to regional programs. Africa's RECs have convening powers that position them as valuable vehicles. That is, they convene meetings of political leaders at the highest level, and these leaders make decisions that are binding on the member states; the member states then regularly report on their performance regarding these decisions. Such meetings provide good platforms for sharing information and best practices. Africa's RECs have established and continue to designate centers of excellence in various areas. COMESA, for instance, has established reference laboratories for animal and plant research in Kenya and Zambia. Designation of centers of excellence for specific aspects of agricultural research will greatly assist specialization within the RECs and put to common use the knowledge from the expertise identified in the region.

Entrepreneurial Leadership

It is not enough for governments to simply reduce the cost of doing business. Fostering agricultural renewal will require governments to function as active facilitators of technological learning. Government actions will need to reflect the entrepreneurial character of the farming community; they too will need to be entrepreneurial. Leadership will also need to be entrepreneurial in character. Moreover, addressing the challenge will require governments to adopt a mission-oriented approach, setting key targets and providing support to farmers to help them meet quantifiable goals. A mission-oriented approach will require greater reliance on executive coordination of diverse departmental activities.

Fostering economic renewal and prosperity in Africa will entail adjustments in the structure and functions of government. More fundamentally, issues related to agricultural innovation must be addressed in an integrated way at the highest possible levels in government. There is therefore a need to strengthen the capacity of presidential offices to integrate science, technology, and innovation in all sustainable agriculture-related aspects of government. Moreover, such offices will also need to play a greater role in fostering interactions between government, business, academia, and civil society. This task requires champions.

One of the key aspects of executive direction is the extent to which leaders are informed about the role of science and innovation in agricultural development. Systematic advice on science and innovation must be included routinely in policymaking. Such advisers must have access to credible scientific or technical information drawing from a diversity of sources, including scientific and engineering academies. In fact, the magnitude of the challenge for regions like Africa is so great that a case could be made for new academies dedicated to agricultural science, technology, and innovation.

Science, technology, and engineering diplomacy has become a critical aspect of international relations. Ministries of foreign affairs in African countries have a responsibility to promote international technology cooperation and to forge strategic alliances on issues related to sustainable agriculture. To effectively carry out this task, foreign ministries need to strengthen their internal capability in science and innovation.

Regional Innovation Communities

Regional integration is a key component of enabling agricultural innovation because it dismantles three barriers to development: "weak national economies; a dependence on importing high-value or finished goods; and a reliance on a small range of low-value primary exports, mainly agriculture and natural resources."[2]

Physical infrastructure creates a challenge for many African countries but also presents an opportunity for the RECs to collaborate on mutually beneficial projects. In many parts of Africa, poor road conditions prevent farmers from getting to markets where they could sell their excess crops profitably. Poor road conditions include the lack of paved roads, the difficulty of finding transportation into market centers, and the high cost of having to pay unofficial road fees to either customs officials or other agents on the roads. These difficulties become more extreme when farmers have to get their crops across international borders to reach markets where sales are profitable.

The inability to sell crops, or being forced to sell them at a loss because of high transportation costs, prevents farmers from making investments that would increase the quantity and quality of their production, since any increase will not add to their own well-being, and the excess crops may go to waste. This is a problem where national governments and regional cooperation offer the best solution. Regional bodies, with representation from all of the concerned countries, are placed to address the needs for better subregional infrastructure and standardization of customs fees at only a few locations.

Having countries come together to address problems of regional trade, particularly including representatives from both the private and public sectors, allows nations to identify and address the barriers to trade. The governments are now working together to address the transportation problem and to standardize a regional system of transport and import taxes that will reduce the cost of transporting goods between nations. This new cooperation will allow the entire region to increase its food security by capitalizing on the different growing seasons in different countries and making products available in all areas for longer periods of time, not just the domestic season. Such cooperation also provides African farmers with access to international markets that they did not have before, since it will allow them to send their goods to international ports, where they can then sell them to other nations.

Regional economic bodies provide a crucial mechanism for standardizing transport procedures and giving farmers a chance to earn money selling their products. This cooperation works best when it happens both between countries and between the private and public sectors. Having multiple actors involved allows for better information, more comprehensive policymaking, and the inclusion of many stakeholders in the decision-making process. Governments and private actors should strengthen their participation in regional bodies and use those groups to address transportation issues, market integration, and infrastructure problems.

Facilitating regional cooperation is emerging as a basis for diversifying economic activities in general and leveraging international partnerships in particular.[3] Many of Africa's individual states are no longer viable economic entities; their future lies in creating trading partnerships with neighboring countries. Indeed, African countries are starting to take economic integration seriously.[4] For example, the re-creation of the EAC is serving not only as a mechanism for creating larger markets but also is promoting peace in the region. Economic asymmetry among countries often is seen as a source of conflict.[5] However, the inherent diversity can serve as an incentive for cooperation.

One of the best examples of regional innovation cooperation was the establishment of the COMESA Innovation Council in April 2013. The Innovation Council applies science and technology advice to foster regional trade and investment by creating a network of scientists and engineers and encouraging innovation by small and medium-sized enterprises. One major goal is to foster collaboration between COMESA's various ministries, especially trade and science and technology. Innovation Council members are drawing from academia and from both the public and private sectors and advise ministers on how best to harness new technologies for development. The Council's work focuses on what COMESA heads of state identify as priorities. Each of COMESA's member states establishes a fund, and COMESA's secretariat establishes a regional fund to disperse grants to match national funds. Finally, the Council supports the regional innovation prizes to be awarded annually. The Council will help COMESA to increase the use of evidence in decisionmaking.

Fostering the Culture of Innovation

When the African Union leaders declared 2014 to be the Year of Agriculture and Food Security in Africa via the Malabo

Declaration in July 2014, they signaled both a serious commitment as well as a new approach to agricultural development.

The Chairperson of the AU Commission, Dr. Nkosazana
Dlamini-Zuma, made it clear that agriculture and agribusiness
are cornerstones of both economic development and empowerment of smallholder farmers, especially women and youth.
She also stressed the important of applying technology and
modern techniques to boost growth in the sector.[6] The overarching objective of the Year of Agriculture is to "consolidate active
commitments toward new priorities, strategies and targets for
achieving results and impacts, with special focus on sustained,
all Africa agriculture-led growth, propelled by stronger, private sector investment and public-private partnerships."[7]

The African Union has committed to reduce poverty by
half, with agricultural growth and transformation as the catalyst, and has promised to put in place the policy, institutional,
and monetary support to make this happen. The African
Union singled out several benchmarks by which to measure
progress: first, agricultural GDP growth would reach at least
6% annually; second, it promised to strengthen public-private
partnerships for key commodity value chains with links to
smallholder farmers; third, it will focus on creating employment opportunities for 30% of the youth in these value chains;
and fourth, it will focus particularly on involving women and
youth in agribusiness opportunities.[8] As has been discussed
throughout this book, Africa possesses a latecomer advantage in adopting existing technology. This is an important
step toward encouraging innovation and entrepreneurship in
agriculture.

Improving the Governance of Innovation

Perhaps the most important recent development is the adoption of the African Union Science, Technology and Innovation
Strategy for Africa 2024 (STISA-2024), a long-term strategy that
supports the African Union's "Agenda 2063." The strategy puts

science, technology, and innovation (STI) firmly at the center of Africa's socioeconomic development and growth agenda. The strategy responds to a growing demand for a knowledge-based economy and evidence-based decisionmaking and policy measures relating to a variety of sectors, including agriculture, energy, environment, health, infrastructure development, mining, security, and water. STISA-2024 builds upon the African Union's Malabo Declaration and will focus on six priority areas: (1) eradication of hunger and achieving food security; (2) prevention and control of diseases; (3) communication (physical and intellectual mobility); (4) protection of African space, including climate change studies; (5) community development, including regional integration and governance and democracy; and (6) wealth creation, including education and human resource development, management of natural resources, and management of water resources.

The strategy is built around three prerequisite pillars that will ensure the successful implementation of STISA-2024: infrastructure, higher technical training, and entrepreneurship. Member states and regions must commit to building capacity in each area to ensure the success of the strategy. Infrastructure development requires upgrading and expanding laboratories to include teaching, engineering, and clinical trials; building teaching hospitals; upgrading ICT equipment and networks; creating innovation spaces; and establishing research and education networks. National governments must create an enabling environment for building infrastructure capacity by facilitating partnerships between scientists and engineers and by training a cadre of people who can ensure that infrastructure is up to par to support STISA-2024. Second, technical competence must be improved. Policymakers need to provide better access to quality secondary and higher education, especially focusing on PhD programs. These programs must promote STI research and innovation as viable career paths. This leads directly to the third pillar: entrepreneurship development. This will require regional coordination and a more systematic approach to stimulate local, national, and regional innovation systems. Policies will have to address technology transfer, knowledge sharing, creation and adaptation of new products, services, business models, and commercialization of research outputs.

That the African Union member states have adopted both Agenda 2063 and STISA-2024 shows significant commitment to improving economic development and well-being throughout the continent. STISA-2024 will go a long way toward promoting agricultural development in particular throughout sub-Saharan Africa.

Promoting a growth-oriented agenda will require adjustments in the structure and functions of government at the regional, national, and local levels. Issues related to science, technology, and innovation must be addressed in an integrated way at the highest possible levels in government. There is therefore a need to strengthen the capacity of presidential offices to integrate science, technology, and innovation in all aspects of government. No African head of state or government currently has a chief scientific adviser.

The intensity and scope of coordination needed to advance agricultural innovation exceeds the mandate of any one ministry or department. As noted elsewhere, Malawi addressed the challenge of coordination failure by presidential control of agricultural responsibilities. The need for high-level or executive coordination of agricultural functions is evident when one takes into account the diverse entities that have direct relevance to any viable programs. Roads are important for agriculture, yet they fall under different ministries that may be more concerned with connecting cities than rural areas. Similarly, ministries responsible for business development may be focusing on urban areas, where there is a perception of short-term returns to investment. The point here is not to enter the debate on the so-called urban bias.[9]

The main point is to highlight the importance of strategic coordination and alignment of the functions of government to reflect contemporary economic needs. Aligning the various organs of government to focus on the strategic areas of economic efforts requires the use of political capital. In nearly all systems of government such political capital is vested in the chief executive of a country, either the president or the prime minister, depending on the prevailing constitutional order. It would follow from this reasoning that presidents or prime ministers should have a critical agricultural coordination role to perform. They can do so by assuming the position of minister or by heading a body charged with agricultural innovation. The same logic also applies for the RECs.

The dominant thinking is to create "science and innovation desks" in the RECs. Such desks will mirror the functions of the science and technology ministries at the national level. It is notable that currently no African leaders are supported by effective mechanisms that provide high-level science, technology, and engineering advice. The absence of such offices (with proper terms of reference, procedures, legislative mandates, and financial resources) hampers the leaders' ability to keep abreast of emerging technological trends and to make effective decisions. Rapid scientific advancement and constant changes in the global knowledge ecology require African leaders at all levels (heads of RECs, presidents, or prime ministers and heads of key local authorities such as states or cities) to start creating institutions for science advice. In 2010 COMESA led the way by adopting decisions on science, technology, and innovation along these lines. Agricultural innovation could be the first beneficiary of informed advice from such bodies.

Bringing science, technology, and engineering to the center of Africa's economic renewal will require more than just political commitment; it will take executive leadership. This challenge requires concept champions, who in this case will be heads of state spearheading the task of shaping their economic policies around science, technology, and innovation. So far, most African countries have failed to develop national policies that demonstrate a sense of focus to help channel emerging technologies into solving developmental problems. They still rely on generic strategies dealing with "poverty alleviation" without serious consideration of the sources of economic growth.

One of the central features of executive guidance is the degree to which political leaders are informed about the role of science, technology, and engineering in development. Advice on science, technology, and innovation must be included routinely in policymaking. An appropriate institutional framework must be created in order for this to happen. Many African cabinet structures are merely a continuation of the colonial model, structured to facilitate the control of local populations rather than to promote economic transformation.

Advisory structures differ across countries. In many countries, science advisers report to the president or prime minister, and national scientific and engineering academies provide political leaders with advice. Whatever structure is adopted, the advising function should have some statutory mandate to advise the highest levels of government. It should have its own operating budget and a budget for funding policy research. The adviser should have access to good and credible scientific or technical information from the government, national academies, and international networks. The advisory processes should be accountable to the public and should be able to gauge public opinion about science, technology, and innovation.

Successful implementation of science, technology, and innovation policy requires civil servants who have the capacity for policy analysis—capacity that most current civil servants lack. Providing civil servants with training in technology management, science policy, and foresight techniques can help integrate science, technology, and innovation advice into decisionmaking. Training diplomats and negotiators in science, technology, and engineering also can increase their ability to discuss technological issues in international forums.

African countries have many opportunities to identify and implement strategic missions or programs that promote growth through investments in infrastructure, technical training, business incubation, and international trade. For example, regional administrators and mayors of cities can work with government, academia, industry, and civil society to design missions aimed at improving the lives of their residents. Universities located in such regions and cities could play key roles as centers of expertise, incubators of businesses, and overall sources of operational outreach to support private and public sector activities. They could play key roles in transferring technology to private firms.[10]

Similar missions could be established in rural areas. These missions would become the organizing framework for fostering institutional interactions that involve technological learning and promote economies of scale. In this context, missions that involve regional integration and interaction should be given priority, especially where they build on local competencies.

This approach can help the international community isolate some critical elements that are necessary when dealing with such a diverse set of problems as conservation of forests, provision of clean drinking water, and improving the conditions of slum dwellers. In all these cases, the first major step is the integration of environmental considerations into development activities.

Reforming the Structures of Innovation Governance

The RECs offer a unique opportunity for Africa to start rethinking the governance of innovation so that the region can propel itself to new frontiers and run its development programs in an enlightened manner that reflects contemporary challenges and opportunities. The focus of improvements in governance structures should be at least in four initial areas: a high-level committee on science, innovation, technology, and engineering; regional science, technology, and engineering academies; an office of science, technology, and innovation; and a graduate school of innovation and regional integration.

Committee on Science, Innovation, Technology, and Engineering

The committee will be a high-level organ of each REC that will report directly to the councils of ministers and presidential summits. Its main functions should be to advise the respective REC on all matters pertaining to science, technology, engineering, and innovation. The functions should include, but not be limited to, regional policies that affect science, technology, engineering, and innovation. It shall also provide scientific and technical information needed to inform and support public policy on regional matters in areas of the competence of the RECs (including economy, infrastructure, health, education, environment, security, and other topics).

For such a body to be effective, it will need to draw its membership from a diversity of sectors, including government, industry, academia, and civil society. The members shall serve for a fixed term, specified at the time of appointment. Within these sectors, representation should reflect the fact that science, technology, engineering, and innovation are not limited to a few ministries or departments but cover the full scope of the proper functioning of society.

The committee should meet as needed to respond to information requests by the chief executive, councils of ministers, or the summits. To meet this challenge, the committee should solicit information from a broad spectrum of stakeholders in the research community, private sector, academia, national research institutes, government departments, local government, development partners, and civil society organizations. The committee's work can be facilitated through working groups or task forces set up to address specific issues.

The committee's work will be supported by the Office of Science, Innovation, Technology, and Engineering, headed by a director who also serves as the chief science, innovation, technology, and engineering adviser to the chief executive. A national analogue of such a committee is India's National Innovation Council, which was established in 2010 by the prime minister to prepare a road map for the country's Decade of Innovation (2010–2020). The aim of the council was to develop an Indian innovation model that focuses on inclusive growth and the creation of institutional networks that can foster inclusive innovation. The council will promote the creation of similar bodies at the sectoral and state levels.[11]

Regional Academies of Science, Innovation, Technology, and Engineering

African countries have in recent years been focusing on creating or strengthening their national academies of science and technology. So far 16 African countries (Cameroon, Egypt, Ethiopia, Ghana, Kenya, Madagascar, Mauritius, Morocco, Mozambique, Nigeria, Senegal, South Africa, Sudan, Tanzania, Uganda, and Zimbabwe) have national academies. There is also the nongovernmental African Academy of Sciences (AAS).

It is notable that, despite Africa's growing emphasis on investing in infrastructure, especially telecommunications, only one African country (South Africa) has an academy devoted to promoting engineering. It is estimated that the continent will need to invest over $93 billion annually to meet its infrastructure requirements. More recently, China has been an important contributor to the expansion of infrastructure facilities in Africa. While China has played in a key role in the speedy construction of infrastructure projects, key questions remain regarding Africa's ability to maintain the facilities after they have been built. A long-term response to the challenge will involve systematic and creative efforts to strengthen the continent's engineering and technological capacity base.

Democratic Republic of Congo joins South Africa as being the second country with a plan to develop an Academy of Engineering. The creation of the new academy is inspired by the need to expand engineering and technology education and practice to meet Africa's infrastructure development shortcomings. The Academy will focus on advancing, inspiring, and celebrating the role of engineering and technology as core foundations for long-term economic transformation, sustainable development, and human well-being. It will mobilize eminent engineers to serve on committees created to provide independent and nonpartisan advice to governments on critical aspects of engineering and technology for development.

Specifically, it will challenge the engineering and technology community to contribute to development; recognize excellence and inspire young generations to engage more actively in engineering and technology activities; provide intellectual leadership and offer advice to governments on the role of engineering in development; build capacity in engineering and technology, while paying particular attention to the role of women; and serve as a role model for other regional and national academies. The Academy will work closely with other institutions such as the African Academy of Sciences and the World Academy of Sciences, as well as national engineering, technological, and scientific academies. Politically, it will work with the African Union and the RECs.

One organization is already addressing the challenge of recognizing excellence among African entrepreneurs. In 2014, the Royal Academy of Engineering established the first Africa Prize for Engineering Innovation, worth £25,000. The Prize also includes six months of mentorship, training, and commercialization for all short-listed candidates from engineering and business development experts. The six-month period is designed to help candidates realize their innovations, scale them up, and commercialize them. The Prize covers all engineering disciplines and was launched to encourage engineers and entrepreneurs specifically to address local challenges by developing scalable solutions. The Prize is one part of the RAE's commitment to help promote engineering throughout the subcontinent in conjunction with several universities as well as the Africa-UK Engineering for Development Partnership.

Most other countries seek to recognize engineers through regular scientific academies, but their criteria for selection tend to focus on publications, rather than practical achievements. A case can be made for the need to expand the role of academies in providing advice on engineering-related investments.

The creation of regional academies of science, innovation, technology, and engineering will go a long way in fostering the integration of the various fields and disciplines so that they can help to foster regional integration and development.[12] The main objectives of such academies would be to bring together leaders of the various regions in science, innovation, technology, and engineering to promote excellence in those fields. Their priorities would be to strengthen capabilities, inspire future generations, inform public debates, and contribute to policy advice.

The fellows of the academies will be elected through a rigorous process following international standards adopted by other academies. Their work and outputs should also follow the same standards used by other academies. The academies should operate on the basis of clear procedures and should operate independently. They may from time to time be asked to conduct studies by the RECs but they should also initiate their own activities, especially in areas such as monitoring scientific, technological, and engineering trends worldwide and keeping the RECs informed about their implications for regional integration and development. Unlike the committee, the academies will operate independently and their advisory functions are only a part of a larger agenda of advancing excellence in science, innovation, technology, and engineering.

If needed, specialized regional academies of agriculture could be created to serve the sector. Such agricultural academies could benefit from partnerships with similar organizations in countries such as China, India, Sweden, and Vietnam. The proposed academies will need to work closely with existing national academies and the AAS.

Office of Science, Innovation, Technology, and Engineering

The RECs will need to create strong offices within their secretariats to address issues related to science, innovation, technology, and engineering. The bulk of the work of such offices will be too coordinate advisory input, as well as serve as a link between the various organs of the RECs and the rest of the world. The head of the office will have two main functions. First, the person will serve as the chief adviser to the various organs of the RECs (through the chief executive). In effect, the person will be the assistant to the chief executive on science, innovation, technology, and engineering. Second, the person will serve as director of the office and will be its representative when dealing with other organizations. In this role the director will be a promoter of science, innovation, technology, and engineering, whereas in the first role the person will serve as an internal adviser.

For such an office to be effective, it will need to be adequately funded and staffed. It can draw from the personnel of other departments, academies, or organizations to perform certain duties. In addition to having adequate resources, the office will need to develop transparent procedures on how it functions and how it relates to other bodies. It is imperative that the functions of the office be restricted to the domain of advice and it should not have operational responsibilities, which belong to the national level.

Too often, policy decisions made in African countries are politically motivated and do not reflect the balance of scientific evidence. Creating a position of chief scientific adviser to presidents, prime ministers, and the chairperson of the African Union would allow African leaders to act strategically and analytically, adopting technologies and innovations when and where it makes sense to do so. Such offices would provide advice on how to capitalize on the renewed interest in agriculture and invest in infrastructure, higher technical training and creation of larger markets.

School of Regional Integration

The need to integrate science and innovation in regional development will require the creation of human capacity needed to manage regional affairs. So far, the RECs rely heavily on personnel originally trained to manage national affairs. There are very few opportunities for training people in regional integration. The newly created COMESA School of Regional Integration could undertake research, professional training, and outreach on how to facilitate regional integration. The creation of the grand free trade area in June 2015 involving 26 African countries has increased the demand for professional training in regional integration. The lessons learned from the implementation of the COMESA School could guide future training within the grand free trade area. The school could fulfill its mission by collaborating with other national and international universities.

The school could focus on providing training on emerging issues such as science and innovation. It can do so through short executive courses, graduate diplomas, and degree programs. There is considerable scope for fostering cooperation between such a school and well-established schools of government and public policy around the world. The theme of regional integration is a nascent field with considerable prospects for growth. For this reason it would not be difficult to promote international partnerships that bring together regional and international expertise.

The school could also serve as depository of knowledge gained in the implementation of regional programs. Staff from the RECs could serve as adjunct faculty and so could join it as full-time professors of the practice of regional integration. The school could also work with universities in the region to transfer knowledge, curricula, and teaching methods to the next generation of development practitioners. The area of agricultural innovation would be ideal for the work of such a school and a network of universities that are part of the regional innovation system.

Funding Innovation

One of the key aspects of technological development is funding. Financing technological innovation should be considered in the wider context of development financing. Lack of political will is often cited as a reason for the low level of financial support for science, technology, and innovation in Africa. But a large part of the problem can be attributed to tax and revenue issues that fall outside the scope of science and technology ministries.[13] For example, instruments such tax credits that have been shown to increase intensity of research and development activities are unlikely to work in policy environments without a well-functioning tax regime.[14] Other instruments such as public procurement can play a key role in stimulating innovation, especially among small and medium-sized enterprises (SMEs).[15]

Currently, Africa does not have adequate and effective mechanisms for providing support to research. Many countries have used a variety of models, including independent funders, such as the National Science Foundation in the United States and the National Research Fund of South Africa. Others have focused on ensuring that development needs guide research funding and, as such, have created specific funding mechanisms under development planning ministries. While this approach is not a substitute for funding to other activities, it distinguishes between measures designed to link technology to the economy from those aimed at creating new knowledge for general learning. What is critical, however, is to design appropriate institutional arrangements and to support funding mechanisms that bring knowledge to bear on development.

Creating incentives for domestic mobilization of financial resources as a basis for leveraging external support would be essential. Other innovations in taxation, already widespread around the world, involve industry-wide levies to fund research, similar to the Malaysian tax mechanism to fund research. Malaysia imposed cesses on rubber, palm oil, and timber to fund the Rubber Research Institute, the Palm Oil Research Institute, and the Forestry Research Institute. A tax on tea helps fund research on and marketing of tea in Sri Lanka. Kenya levies a tax on its tea, coffee, and sugar industries, for example, to support the Tea Research Foundation, the Coffee Research Foundation, and the Kenya Sugar Board.

These initiatives could be restructured to create a funding pool to cover common areas. Reforming tax laws is an essential element in the proposed strategy. Private individuals and corporations need targeted tax incentives to contribute to research funds and other technology-related charitable activities. This instrument for supporting public welfare activities is now widely used in developing countries. It arises partly because of the lack of experience in managing charitable organizations and partly because of the reluctance of finance ministries to grant tax exemptions, fearing erosion of their revenue base.

The enactment of a foundation law that provides tax and other incentives to contributions to public interest activities, such as research, education, health, and cultural development, would promote social welfare in general and economic growth in particular. Other countries are looking into using national lotteries as a source of funding for technological development. Taxes on imports could also be levied to finance innovation activities, although the World Trade Organization may object to them. Another possibility is to impose a tax of 0.05% or 0.1% of the turnover of African capital markets to establish a global research and development fund, as an incentive for them to contribute to sustainable development.

Other initiatives could simply involve restructuring and redefining public expenditure. By integrating research and development activities into infrastructure development, for example, African governments could relax the public expenditure constraints imposed by sectoral budgetary caps. Such a strategy has the potential to unlock substantial funds for research and development in priority areas. But this strategy requires a shift in the budgetary philosophy of the international financial institutions to recognize public expenditures on research and development as key to building capabilities for economic growth.

Financing is probably one of the most contentious issues in the history of higher education. The perceived high cost of running institutions of higher learning has contributed to the dominant focus on primary education in African countries. But this policy has prevented leaders from exploring avenues for supporting higher technical education.

Indeed, African countries such as Uganda and Nigeria have considered new funding measures, including directed government scholarships and lower tuition for students going into the sciences. Other long-term measures include providing tax incentives to private individuals and firms that create and run technical institutes on the basis of agreed government policy. Africa has barely begun to utilize this method as a way to extend higher technical education to a wider section of society. Mining companies, for example, could support training in the geosciences. Similarly, agricultural enterprises could help create capacity in business.

Institutions created by private enterprises can also benefit from resident expertise. Governments, on the other hand, will need to formulate policies that allow private sector staff to serve as faculty and instructors in these institutions. Such programs also would provide opportunities for students to interact with practitioners in addition to the regular faculty.

Much of the socially responsible investment made by private enterprises in Africa could be better used to strengthen the continent's technical skill base. Additional sources of support could include the conversion of the philanthropic arms of various private enterprises into technical colleges located in Africa.

Governmental and other support will be needed to rehabilitate and develop university infrastructures, especially information and communications facilities, to help them join the global knowledge community and network with others around the world. Such links will also help universities tap into their experts outside the country. Higher technical education should also be expanded by creating universities under line ministries, as pioneered by telecom universities such as the Nile University (Egypt), the Kenya Multimedia University, and the Ghana Telecoms University College. Other line ministry institutions, such as the Digital Bridge Institute in Nigeria, are also considering becoming experiential universities with strong links with the private sector.

Governments and philanthropic donors could drive innovation through a new kind of technology contest.[16] One approach is to offer proportional "prize rewards" that would modify the traditional winner-take-all approach by dividing available funds among multiple winners in proportion to measured achievement.[17] This approach would provide a royalty-like payment for incremental success.[18]

Promoting innovation for African farmers has proven especially challenging, due to a wide variety of technological and institutional obstacles. A proportional-prize approach is particularly suited to help meet the needs of African farmers. For that purpose, a specific method should be devised to implement prize rewards, to recognize and reward value creation from new technologies after their adoption by African farmers.

In summary, the effectiveness of innovation funding depends on choosing the right instrument for each situation—and perhaps, in some situations, developing a new instrument that is specifically suited to the task. Prizes are distinctive in that they are additional and temporary sources of funding, they are used when needed to elicit additional effort, and they can reveal the most successful approaches for reaching a particular goal. For this reason, a relatively small amount of funding in a well-designed prize program can help guide a much larger flow of other funds, complementing rather than replacing other institutional arrangements.

Available evidence suggests that investments in agricultural research require long-term sustained commitment. This is mainly because of the long time lags associated with such investments, ranging from 15 to 30 years, taking into account the early phases of research.[19] Part of the time lag, especially in areas such as biotechnology, is accounted for by delays in regulatory approvals or the high cost of regulation. This is true even in cases where products have already been approved and are in use in technology pioneering countries.[20] These long time lags are also an expression of the fact that the economic systems co-evolve with technology and the process involves adjustments in existing institutions.[21]

Joining the Global Knowledge Ecology

Leveraging Africa's Diasporas

Much of the technological foundation needed to stimulate African development is based on ideas in the public domain (where property rights have expired). The challenge lies in finding ways to forge viable technology alliances.[22] In this regard, intellectual property offices are viewed as important sources of information needed for laying the basis for technological innovation.[23] While intellectual property protection is perceived as a barrier to innovation, the challenges facing Africa lie more in the need to build the requisite human and institutional capability to use existing technologies. Much of this can be achieved though collaboration with leading research firms and product development.[24] This argument may not hold in regard to emerging fields such as genomics and nanotechnology.

One of the concerns raised about investing in technical training in African countries is the migration of skilled manpower to industrialized countries.[25] The World Bank has estimated that although skilled workers account for just 4% of the sub-Saharan labor force, they represent some 40% of its migrants.[26] Such studies tend to focus on policies that seek to curb the so-called brain drain.[27] But they miss the point. The real policy challenge for African countries is figuring out how to tap the expertise of those who migrate and upgrade their skills while out of the country, rather than engaging in futile efforts to stall international migration.[28] The most notable case is the Taiwanese diaspora, which played a crucial role in developing the country's electronics industry.[29] This was a genuine partnership involving the mobility of skills and capital.

Countries such as India have studied this model and have come to the conclusion that one way to harness the expertise is to create a new generation of "universities for innovation" that will seek to foster the translation of research into commercial products. In 2010 India unveiled a draft law that will provide for the establishment of such universities. The law grew out India's National Knowledge Commission, a high-level advisory body to the prime minister aimed at transforming the country into a knowledge economy.[30]

A number of countries have adopted policy measures aimed at attracting expatriates to participate in the economies of their countries of origin. They are relying on the forces of globalization such as connectivity, mobility, and interdependence to promote the use of the diaspora as a source of input into national technological and business programs. These measures include investment conferences, the creation of rosters of experts, and direct appeals by national leaders. It is notable that expatriates are like any other professionals and are unlikely to be engaged in their countries of origin without the appropriate incentives. Policies or practices that assume that these individuals owe something to their countries of origin are unlikely to work.

Considerable effort needs to be put into fostering an atmosphere of trust between the expatriates and local communities. In addition, working from a common objective is critical, as illustrated in the case of the reconstruction of Somaliland. In this inspirational example, those involved in the Somaliland diaspora were able to invoke their competence, networks, and access to capital to establish the University of Hargeisa, which has already played a critical role in building the human resource base needed for economic development. The achievement is even more illustrative when one considers the fact that the university was built after the collapse of Somalia.[31] Ashesi University in Accra, Ghana, is another example of the role that the diaspora can play in local development efforts. In 2001, Patrick Awuah returned to Ghana after two decades in the United States. Ashesi University began with 30 students in a rented building with an admissions office, a library, a classroom, a computer lab, and a cafeteria. By 2011 it had a permanent campus in Berekuso comprising 9 buildings and 500 students. Awuah's vision was to create a liberal arts university that would fill a void in higher education by creating entrepreneurial leaders and teaching critical thinking and problem-solving skills, with a focus on practical experience. A new engineering building is being planned that will educate engineer entrepreneurs who will address infrastructure and other local needs. Similar efforts involving the Somali diaspora in collaboration with King's College Hospital in London have contributed significantly to the health care sector in Somaliland.[32] There are important lessons in this case that can inform the rest of Africa. The initial departure of nationals to acquire knowledge and skills in other countries represents a process of upgrading their skills and knowledge through further training. But returning home without adequate opportunities to deploy the knowledge earned may represent the ultimate brain drain. A study of Sri Lankan scientists in diaspora has shown that further studies "was the major reason for emigration, followed by better career prospects. Engineering was the most common specialization, followed by chemistry, agricultural sciences and microbiology/ biotechnology/molecular biology. If their demands are adequately met, the majority of the expatriates were willing to return to Sri Lanka."[33]

Science and Innovation Diplomacy

The area of science, technology, and engineering diplomacy has become a critical aspect of international relations.[34] Science is gaining in prominence as a tool fostering cooperation and resolving disputes among nations.[35] Much of the leadership is provided by industrialized countries. For example, the United States has launched a program of science envoys, which is adding a new dimension to US foreign policy.[36] This diplomatic innovation is likely to raise awareness of the importance of science, technology, and engineering in African countries.

Ministries of foreign affairs have a responsibility in promoting international technology cooperation and forging strategic alliances. To effectively carry out this mandate, these ministries need to strengthen their internal capability in science, technology, and innovation. To this end, they will need to create offices dealing specifically with science, technology, and engineering, working in close cooperation with other relevant ministries, industry, academia, and civil society. Such offices could also be responsible for engaging and coordinating expatriates in Africa's technology development programs.

There has been growing uncertainty over the viability of traditional development cooperation models. This has inspired the emergence of new technology alliances involving the more advanced developing countries.[37] For example, India, Brazil, and South Africa have launched a technology alliance that will focus on finding solutions to agricultural, health, and environmental challenges. In addition, more developing countries are entering into bilateral partnerships to develop new technologies. Individual countries such as China and Brazil are also starting to forge separate technology-related alliances with African countries. Brazil, for example, is increasing its cooperation with African countries in agriculture and other fields.[38] In addition to establishing a branch of the Brazilian Agricultural Research Corporation (EMBRAPA) in Ghana, the country has also created a tropical agricultural research institute at home to foster cooperation with African countries.

Significant experiments are under way around the world to make effective use of citizens with scientific expertise who are working abroad. The UK consulate in Boston is engaged in a truly pioneering effort to advance science, technology, and engineering diplomacy. Unlike other consulates dealing with regular visa and citizenship issues, the consulate is devoted to promoting science, technology, and engineering cooperation between the United Kingdom and the United States while also addressing major global challenges such as climate change and international conflict.

In addition to Harvard University and MIT, the Boston area is home to more than 60 other universities and colleges, making it the de facto intellectual capital of the world. Switzerland has also converted part of its consulate in Boston into a focal point for interactions between Swiss experts in the United States and their counterparts at home. Swissnex was created in recognition of the importance of having liaisons in the area, which many consider the world's leading knowledge center, especially in the life sciences. These developments are changing the way in which governments envision the traditional role of science attachés, with many giving them more strategic roles.[39]

In another innovative example, the National University of Singapore has established a college at the University of Pennsylvania to focus on biotechnology and entrepreneurship. The complementary Singapore-Philadelphia Innovators' Network (SPIN) serves as a channel and link for entrepreneurs, investors, and advisers in the Greater Philadelphia region and

Singapore. The organization seeks to create opportunities for international collaboration and partnerships in the area.

India, on the other hand, has introduced changes in its immigration policy, targeting its citizens working abroad in scientific fields to strengthen their participation in development at home. Such approaches can be adopted by other developing countries, where the need to forge international technology partnerships may be even higher, provided there are institutional mechanisms to facilitate such engagements.[40] The old-fashioned metaphor of the "brain drain" should to be replaced by a new view of "global knowledge flows."[41]

But even more important is the emerging interest among industrialized countries to reshape their development cooperation strategies to reflect the role of science, technology, and innovation in development. The UK Department for International Development (DFID) took the lead in appointing a chief scientist to help provide advice to the government on the role of innovation in international development, a decision that was later emulated by USAID.[42] Japan has launched a program on science and technology diplomacy that seeks to foster cooperation with developing countries on the basis of its scientific and technological capabilities.[43] Similarly, the United States has initiated efforts to place science, technology, and innovation at the center of its development cooperation activities.[44] The initiative will be implemented through USAID as part of the larger science and technology diplomacy agenda of the US government.[45]

South Korea is another industrialized country that is considering adopting a science and innovation approach to development cooperation. These trends might inspire previous champions of development, such as Sweden, to consider revamping their cooperation programs. These efforts are going to be reinforced by the rise of new development cooperation models in emerging economies such India, Brazil, and China. India is already using its strength in space science to partner with African countries. Brazil, on the other hand, positioning itself as a leading player in agricultural cooperation with African countries, is seeking to expand the activities of the Brazilian Development Cooperation Agency.

China's cooperation with Africa is increasingly placing emphasis on science, technology, and engineering. It is a partner in 100 joint demonstration projects and postdoctoral fellowships, which include donations of nearly US$22,000 worth of scientific equipment. China has offered to build 50 schools and train 1,500 teachers and principals, as well as training 20,000 professionals by 2012. The country will increase its demonstration centers in Africa to 20, send 50 technical teams to the continent, and train 2,000 African agricultural personnel. Admittedly, these numbers are modest given the magnitude of the challenge, but they show a shift toward using science, technology, and engineering as tools for development cooperation.[46]

Harmonization of Market Regional Integration

When the heads of state and government of the Common Market for Eastern and Southern Africa, the East African Community, and the Southern African Development Community met in Kampala on October 22, 2008, they conveyed in their communiqué a palpable sense of urgency in calling for the establishment of a single free trade area covering the 26 countries of COMESA, EAC, and SADC. These are 26 of the 55 countries that make up the continent of Africa. The political leaders requested the secretariats of the three organizations to prepare all the legal documents necessary for establishing the single free trade area (FTA) and to clearly identify the steps required (paragraph 14 of the communiqué). In November 2009 the chief executives of the three secretariats cleared the documents for transmission to the member states for consideration in preparing for the next meeting of the Tripartite Summit. The main document is the draft agreement establishing the Tripartite Free Trade, with its 14 annexes covering various complementary areas that are necessary for effective functioning of a regional market. There is a report explaining the approach and the modalities. The main proposal is to establish the FTA on a tariff-free, quota-free, exemption-free basis by simply combining the existing FTAs of COMESA, EAC, and SADC. It was expected that by 2012, none of these FTAs would have any exemptions or sensitive lists. However, there is a possibility that a few countries might wish to consider maintaining a few sensitive products in trading with some big partners, and for this reason, provision has been made for the possibility of a country requesting permission to maintain some sensitive products for a specified period of time.

To have an effective tripartite FTA, various complementary areas have been included. The FTA covers the promotion of customs cooperation and trade facilitation; the harmonization and coordination of industrial and health standards; the combating of unfair trade practices and import surges; the use of peaceful and agreed dispute settlement mechanisms; the application of simple and straightforward rules of origin that recognize inland transport costs as part of the value added in production; and the relaxation of restrictions on the movement of businesspersons, taking into account certain sensitivities.

It also seeks to liberalize certain priority service sectors on the basis of existing programs; to promote value addition and transformation of the region into a knowledge-based economy through a balanced use of intellectual property rights and information and communications technology; and to develop the cultural industries. The tripartite FTA is underpinned by robust infrastructure programs designed to consolidate the regional market through interconnectivity (facilitated, for instance, by all modes of transport and telecommunications) and to promote competitiveness (for instance, through adequate supplies of energy).

Member states adopted an evolutionary road map, building on lessons learned in the integration of existing trade areas. The grand trade area builds on market integration, infrastructure expansion, and industrial development. The learning-based approach involves member state studies, consultations, document adoption, progress monitoring, impact evaluation, and milestone adjustment. The institutional framework for the process was adopted, and formal negotiations were launched in June 2011 in South Africa. It includes a summit, council of ministers, sectoral ministerial committees, and a negotiations forum. The first phase of negotiations (lasting up to one year) covers trade in goods (including tariff liberalization, rules of origin, customs cooperation, non-tariff barriers, trade remedies, sanitary and phytosanitary measures, technical barriers to trade, and dispute settlement). The second phase (lasting up to five years) covers trade-related issues (including trade in services, intellectual property rights, competition policy, trade development, and competitiveness). Phase One of the Tripartite FTA Agreement was officially concluded at the Third Summit of Heads of State and Government in December 2014 with the signing of the "Declaration on the Conclusion of Negotiations on Phase One—Trade in Goods." The final phase of negotiations resulted in the adoption of the grand free trade area in June 2015 in Cairo. The $1.3 trillion free trade area paved the way for other regional integration efforts.

The main benefit of the Tripartite FTA is that it will be a much larger market, with a single economic space, than any one of the three regional economic communities and as such will be more attractive to investment and large-scale production. Estimates are that exports among the 26 tripartite countries increased from US$7 billion in 2000 to US$30 billion in 2010, and imports grew from US$9 billion in 2000 to US$40 billion in 2010. This phenomenal increase was in large measure spurred by the free trade area initiatives of the three organizations. Strong trade performance, when well designed—for instance, by promoting small and medium-scale enterprises that produce goods or services—can assist the achievement of the core objectives of eradicating poverty and hunger, promoting social justice and public health, and supporting all-around human development. Besides, the tripartite economic space will help to address some current challenges resulting from multiple membership by advancing the ongoing harmonization and coordination initiatives of the three organizations to achieve convergence of programs and activities, and in this way will greatly contribute to the continental integration process. And as they say, the more we trade with each other, the less likely we are to engage in war, for our swords will be plowshares.

Harmonization of Regulations

The need to enhance the use of science, technology, and engineering in development comes with new risks. Africa has not had a favorable history with new technologies. Much of its history has been associated with the use of technology as tools of domination or extraction.[47] The general mood of skepticism toward technology and the long history of exclusion created a political atmosphere that focused excessively on the risks of new technologies. This outlook has been changing quite radically as Africa enters a new era in which the benefits of new technologies to society are widely evident. These trends are reinforced by political shifts that encourage great social inclusion.[48] It is therefore important to examine the management of technological risks in the wider social context, even if the risk assessment tools that are applied are technical.

The risks associated with new technologies need to be reviewed on a case-by-case basis and should be compared with base scenarios, many of which would include risks of their own. In other words, deciding not to adopt new technologies may only compound the risks associated with the status quo. Such an approach would make risk management a knowledge-based process. This would in turn limit the impact of popular tendencies that prejudge the risks of technologies based on their ownership or newness.

Ownership and newness may have implications for technological risks, but they are not the only factors that need to be considered. Fundamentally, decisions on technological risks should take into account the impacts of incumbent technologies or the absence of any technological solutions to problems.

One of the challenges facing African countries is the burden of managing technological risks through highly fragmented systems in contiguous countries. The growing integration of African countries through the RECs offers opportunities to rationalize and harmonize their regulatory activities related to agricultural innovation.[49]

This is already happening in the medical sector. The African Medicines Regulatory Harmonization (AMRH) initiative was established to assist African countries and regions to respond to the challenges posed by medicine registration, as an important but neglected area of medicine access. It seeks to support African Regional Economic Communities and countries in harmonizing medicine registration.

COMESA, in collaboration with the Association for Strengthening Agricultural Research in Eastern and Central Africa (ASARECA) and other implementing partners, has engaged in the development of regionally harmonized policies and guidelines through the Regional Agricultural Biotechnology and Bio-safety Policy in Eastern and Southern Africa initiative since 2003. The COMESA harmonization agenda—now implemented through its specialized agency, the Alliance for Commodity Trade in Eastern and Central Africa—was initiated to provide mechanisms for wise and responsible use of genetically modified organisms in commercial planting, trade, and emergency food assistance.

COMESA, within its mandate of regional economic integration, recognizes the need to support member states in resolving non-tariff barriers that constrain markets and stifle the integration of food products into regional and global value chains, as an innovative strategy to promote market access to regional and international trade.

Such systems are vital to assuring the quality, safety, and efficacy of locally manufactured products and their positive contribution to public health. Moreover, the success of domestic production will partly depend on intra-regional and intra-continental trade to create viable market sizes. Currently, trade in pharmaceuticals is hampered by disparate regulatory systems, which create technical barriers to the free movement of products manufactured in Africa (and beyond)—and have negative consequences for timely patient access to high-quality essential medicines.

However, the implementation of these policies and plans has suffered from a lack of financial and technical resources and has not progressed significantly. Moreover, RECs continue to work largely in isolation. Coordination is needed to avoid duplication of effort and ensure consistent approaches, especially given that more than three-quarters of African countries belong to two or more RECs.

Conclusion

Promoting a growth-oriented agenda will entail adjustments in the structure and functions of government. More fundamentally, issues related to science, technology, and innovation will need to be addressed in an integrated way at the highest level possible in government. Bringing science, technology, and engineering to the center of Africa's economic renewal will require more than just political commitment; it will take executive leadership. This challenge requires concept champions who in this case will be heads of state spearheading the task of shaping their economic policies around science, technology, and innovation.

So far, most African countries have not developed national policies that demonstrate a sense of focus to help channel emerging technologies into solving developmental problems. They still rely on generic strategies dealing with "poverty alleviation," without serious consideration of the sources of economic growth. There are signs of hope, though. NEPAD's Ministerial Forum on Science and Technology played a key role in raising awareness among Africa's leaders of the role of science, technology, and engineering in economic growth.

An illustration of this effort is the decision of the African Union and NEPAD to set up a high-level African Panel on Modern Biotechnology to advise the African Union, its member states, and its various organs on current and emerging issues associated with the development and use of biotechnology. The panel's goal is to provide the African Union and NEPAD with independent and strategic advice on biotechnology and its implications for agriculture, health, and the environment. It focuses on intra-regional and international regulation of the development and application of genetic modification and its products.

Regarding food security in particular, Africa's RECs have tried to develop regional policies and programs to allow member states to work collectively. ECOWAS and COMESA, for instance, building on CAADP, have elaborated regional compacts to guide member states in formulating their national CAADP compacts. This comes at a time when experience from the seven COMESA and 17 ECOWAS national CAADP compacts thus far concluded show that there are key cross-border challenges that will require a regional approach. In addition to 27 African countries that had signed CAADP compacts by 2011, ECOWAS has developed a regional plan to implement CAADP. This has reinforced the need for regional approaches to agricultural development.

Thus, there is a need for a larger regional market to support investment in agricultural products and the harmonization of standards across the region. This will help address challenges such as sanitary and phytosanitary measures that affect the quality and marketability of agricultural products; management of trans-boundary resources such as water bodies and forests; building of regional infrastructure; promotion of collaborative research; monitoring of key commitments of member states, particularly the one on earmarking 10% of the national budget for the agriculture sector.

For Africa to effectively integrate into the global value chain, it needs to embrace global quality standards starting inside the farm gate. One organization is leading the way, not just in defining these standards, but in offering training and certification in the fields of Africa. GlobalG.A.P. is an independent certification system for Good Agricultural Practice (GAP), founded by members of the Euro-Retailer Produce Working Group in 1997. The standards cover sustainable production methods, food safety, animal and worker welfare, plant propagation materials, and compound feed. The harmonized certification simplifies the audit process for producers. GLOBALG.A.P. is now the world's premiere farm assurance program meeting consumer safety needs in more than 100 countries.

A key pillar of CAADP relates to agricultural research and innovation. Africa's RECs have a critical role to play under this pillar, through supporting regional research networks and prioritizing agricultural research in regional policies. Experience sharing at the regional level, and the resulting research communities, will greatly enrich individual research.

Governing agricultural innovation is a complex activity requiring high-level coordination to ensure that all the key functions of government are focused on advancing agricultural innovation. Governing agricultural transformation provides African leaders the opportunity to build up the capacity necessary to become innovation states and surpass the limits of the entrepreneurial state, which is usually focused on promoting global competitiveness. An innovation state has the added challenge of addressing more complex challenges such as inclusive growth and sustainable development.

9

PLOWING AHEAD

A new economic vision for Africa's agricultural transformation—articulated at the highest level of government through Africa's Regional Economic Communities— should be guided by new conceptual frameworks that define the continent as a learning society. This shift will entail placing policy emphasis on emerging opportunities such as renewing infrastructure, building human capabilities, stimulating agribusiness development, and increasing participation in the global economy. It also requires an appreciation of emerging challenges, such as climate change, and the ways in which these challenges may influence current and future economic strategies.

Climate Change, Agriculture, and Economy

As Africa prepares to address its agricultural challenges, it is now confronted with new threats arising from climate change. Agricultural innovation will now have to be done in the context of a more uncertain world in which activities such as plant and animal breeding will need to be anticipatory.[1] According to the World Bank, warming "of 2°C could result in a 4 to 5% permanent reduction in annual income per capita in Africa and South Asia, as opposed to minimal losses in high-income countries and a global average GDP loss of about 1%. These losses would be driven by impacts in agriculture, a sector important to the economies of both Africa and South Asia."[2] Sub-Saharan Africa is dominated by fragile ecosystems. Nearly 75% of its surface area is dry land or desert. This makes the continent highly vulnerable to droughts and floods. Traditional cultures cope with such fragility through migration. But such migration has now become a source of insecurity in parts of Africa. Long-term responses will require changes in agricultural production systems.[3]

The continent's economies are also highly dependent on natural resources. Nearly 80% of Africa's energy comes from biomass, and over 30% of its GDP comes from rain-fed agriculture, which supports 70% of the population. Stress is already being felt in critical resources such as water supply. Today, 20 African countries experience severe water scarcity, and another 12 will endure similar water shortages in the next 25 years. Economic growth in regional hubs is now being curtailed by water shortages.

The drying up of Lake Chad (shared by Nigeria, Chad, Cameroon, and Niger) is a grim reminder that rapid ecological change can undermine the pursuit for prosperity. The lake's area has decreased by 80% over the last 30 years, with catastrophic impacts to local communities. Uncertainty over water supply affects decisions in other areas, such as hydropower, agriculture, urban development, and overall land-use planning. This is happening at a time when Africa needs to switch to low-carbon energy sources.

Technological innovation will be essential for enabling agriculture to adapt to a different climate. Meeting the dual challenges of expanding prosperity and adapting to climate change will require greater investment in the generation and diffusion of new technologies. Basic inputs such as provision of meteorological data could help farmers to adapt to climate change by choosing optimal planting dates.[4] The task ahead for policymakers will be to design climate-smart innovation systems that shift economies toward low-carbon pathways.

Economic development is an evolutionary process that involves adaptation to changing economic environments.

Technological innovation is implicitly recognized as a key aspect of adaptation to climate change. For example, the Intergovernmental Panel on Climate Change (IPCC) defines adaptation as "[a]djustment in natural or human systems in response to actual or expected climatic stimuli or their effects, which moderates harm or exploits beneficial opportunities."[5] It views the requisite adaptive capacity as the ability "to moderate potential damages, to take advantage of opportunities, or to cope with the consequences."[6] Technological innovation is used in society in a congruent way to respond to economic uncertainties. What is therefore needed is to develop analytical and operational frameworks that would make it easier to incorporate adaptation to climate change in innovation strategies that aim to expand prosperity.

Innovation systems are understood to mean the interactive process involving key actors in government, academia, industry, and civil society to produce and diffuse economically useful knowledge into the economy. The key elements of innovation include the generation of a variety of avenues, their selection by the market environment, and the emergence of robust socioeconomic systems. This concept can be applied to adaption to climate change in five critical areas: managing natural resources; designing physical infrastructure; building human capital, especially in the technical fields; fostering entrepreneurial activities; and governing adaptation as a process of innovation.

Economic development is largely a process by which knowledge is applied to convert natural resources into goods and services. The conservation of nature's variety is therefore a critical aspect of leaving options open for future development. Ideas such as "sustainable development" have captured the importance of incorporating the needs of future generations into our actions. Adaptive strategies will therefore need to start with improved understanding of the natural resource base. Recent advances in earth observation and related geo-spatial science and technology have considerably increased the capacity of society to improve its capabilities for natural resource management. But improved understanding is only the first step.

The anticipated disruptive nature of climate change will demand increased access to diverse natural assets such as ge-netic resources for use in agriculture, forestry, aquaculture, and other productive activities. For example, the anticipated changes in the growing season of various crops will require intensified crop breeding.[7] But such breeding programs will presuppose not only knowledge of existing practices but also the conservation of a wider pool of genetic resources of ex-isting crops and breeds and their wild relatives to cope with shifts in agricultural production potential.[8] This can be done through measures such as seed banks, zoos, and protected areas. Large parts of Africa may have to switch from crop pro-duction to livestock breeding.[9] Others may also have to change from cultivating cereals to growing fruits and vegetables, as projected in other regions of the world.[10] Other measures will include developing migration corridors to facilitate ecosystem integrity and protect human health—through surveillance and early warning systems.

Such conservation efforts will also require innovation in regional institutional coordination, expanded perspectives of space and time, and the incorporation of climate change scenarios in economic development strategies.[11] Building robust economies requires the conservation of nature's va-riety. These efforts will need to be accompanied by greater investment in the generation of knowledge associated with natural resources. Advances in information and communica-tion capabilities will help the international community to col-lect, store, and exchange local knowledge in ways that were not possible in the past. The sequencing of genomes provides added capacity for selective breeding of crops and livestock suited to diverse ecologies. Technological advancement is therefore helping to augment nature's diversity and expand adaptive capabilities.

Climate change is likely to affect existing infrastructure in ways that are not easy to predict. For example, road networks and energy sources in low-lying areas are likely to be affected by sea-level rise. A recent study of Tangier Bay in Morocco projects that sea-level rise will have a significant impact on the region's infrastructure facilities, such as coastline protection, the port, railway lines, and the industrial base in general.[12]

Studies of future disruptions in transportation systems reveal great uncertainties in impact, depending on geographical location.[13] These uncertainties are likely to influence not only investment decisions but also the design of transportation systems. Similarly, uncertainty over water supply is emerging as a major concern, demanding not only integrated management strategies but also improved use of water-related technologies.

Other measures include the need to enhance water supply—such as linking reservoirs, building new holding capacity in reservoirs, and injecting early snowmelt into groundwater reservoirs. Similarly, coastal areas need to be protected with natural vegetation or seawalls. In effect, greater technical knowledge and engineering capabilities will need to be marshaled to design future infrastructure in light of climate change.[14] This includes the use of new materials arising from advances in fields such as nanotechnology.

Protecting human populations from the risks of climate change should be one of the first steps in seeking to adapt to climate change. Concern over human health can compound the sense of uncertainty and can undermine other adaptive capabilities. Indeed, the first step in building resilience is to protect human populations against disease.[15] Many of the responses needed to adapt health systems to climate change will involve practical options that rely on existing knowledge.[16]

Others, however, will require the generation of new knowledge. Advances in fields such as genomics are making it possible to design new diagnostic tools that can be used to detect the emergence of new infectious diseases. These tools, combined with advances in communications technologies, can be used to detect emerging trends in health and to provide health workers with early opportunities to intervene. Furthermore, convergence in technological systems is transforming the medical field. For example, the advent of handheld diagnostic devices and video-mediated consultation are expanding the prospects of telemedicine, making it easier for isolated communities to be connected to the global health infrastructure.[17] Personalized diagnostics is also becoming a reality.[18]

Adapting to climate change will require significant upgrading of the knowledge base of society. Past failure to adapt from incidences of drought is partly explained by the lack of the necessary technical knowledge needed to identify trends and design responses.[19] The role of technical education in economic development is becoming increasingly obvious. Similarly, responding to the challenges of climate will require considerable investment in the use of technical knowledge at all levels in society.

One of the most interesting trends is the recognition of the role of universities as agents of regional economic renewal.[20] Knowledge generated in centralized urban universities is not readily transferred to regions within countries. As a result, there is growing interest in decentralizing the university system itself.[21] The decentralization of technical knowledge to a variety of local institutions will play a key role in enhancing local innovation systems that can help to spread prosperity through climate-smart strategies.

The ability to adapt to climate change will not come without expertise. But expertise is insufficient unless it is used to identify, assess, and take advantage of emerging opportunities through the creation of new institutions or the upgrading of existing ones. Such entrepreneurial acts are essential for both economic development and adaptation to climate change.

Economic diversification is critical in strengthening the capacity of local communities to adapt to climate change.

For example, research on artisan fisheries has shown that the poorest people are not usually the ones who find it hardest to adapt to environmental shocks. Rather, it is often those who have become locked into overly specialized fishery practices who are most adversely affected.[22] Technological innovation aimed at promoting diversification of entrepreneurial activities would not only help to improve economic welfare, but also help enhance the adaptive capabilities of local communities. But such diversification will need to be complemented by other measures, such as flexibility, reciprocity, redundancy, and buffer stocks.[23]

Promoting prosperity and creating robust economies that can adapt to climate change should be a central concern of leaders around the world. Political turmoil in parts of Africa is linked to recent climate events.[24] The implications of climate change for governance, especially in fragile states, have yet to receive attention.[25] Governments will need to give priority to adaption to climate change as part of their economic development strategies. But they will also need to adopt approaches that empower local communities to strengthen their adaptive capabilities. Traditional governance practices, such as participation, will need to be complemented by additional measures that enhance social capital.[26]

The importance of technological innovation in adaptation strategies needs to be reflected in economic governance strategies at all levels. It appears easier to reflect these considerations in national economic policies. However, similar approaches also need to be integrated into global climate governance strategies, especially through the adoption of technology-oriented agreements.[27]

On the whole, an innovation-oriented approach to climate change adaptation will need to focus largely on expanding the adaptive capacity of society though the conservation of nature's variety, construction of robust infrastructure, enhancement of human capabilities, and promotion of entrepreneurship. Fundamentally, the ability to adapt to climate change will possibly be the greatest test of our capacity for social learning. Regional integration will provide greater flexibility and geographical space for such learning. Furthermore, promoting local innovation as part of regional strategies will contribute to the emergence of more integrated farming systems.[28]

Throughout, this book has highlighted the role that RECs can have as a collective framework for harnessing national initiatives and sharing best practices drawn from the region and beyond. Africa's RECs, as well as the African Union at the continental level, have programs for food security and for science, technology, and engineering. The challenge, as highlighted, relates to putting existing knowledge within the region and beyond to the service of the people of Africa on the ground, through clear political and intellectual leadership and an effective role for innovators. Further, there is the challenge of how best to utilize existing regional policymaking and monitoring and evaluation structures in promoting innovation and tackling the challenges of food security.

The global economic crisis and rising food prices forced the international community to review its outlook for human welfare and prosperity. Much of the concern on how to foster development and prosperity in Africa reflects the consequences of recent neglect of sustainable agriculture and infrastructure as drivers of development. Sustainable agriculture has, through the ages, served as the driving force behind national development. In fact, it has been a historical practice to use returns from investment in sustainable agriculture to stimulate industrial development. Restoring it to its right place in the development process will require world leaders to take a number of bold steps.

Science and innovation have always been the key forces behind agricultural growth in particular and economic transformation in general. More specifically, the ability to add value to agricultural produce via the application of scientific knowledge to entrepreneurial activities stands out as one of the most important lessons of economic history. Reshaping sustainable agriculture as a dynamic, innovative, and rewarding sector in Africa will require world leaders to launch new initiatives that include the following strategic elements.

Bold leadership, driven by heads of state in Africa and supported by those of developed and emerging economies, is needed to recognize the real value of sustainable agriculture in the economy of Africa. High-level leadership is essential for establishing national visions for sustainable agriculture and rural development, championing of specific missions for lifting productivity and nutritional levels with quantifiable targets, and the engagement of cross-sectoral ministries in what is a multisector process.

Sustainable agriculture needs to be recognized as a knowledge-intensive productive sector that is mainly carried out in the informal private economy. The agricultural innovation system must link the public and private sectors and create close interactions among government, academia, business, and civil society. Reforms will need to be introduced in knowledge-based institutions to integrate research, university teaching, farmers' extension, and professional training, bringing them into direct involvement with the production and commercialization of products.

Policies have to urgently address affordable access to communication services for people to use in their everyday lives, as well as broadband Internet connectivity for centers of learning such as universities and technical colleges. This is vital to access knowledge and trigger local innovations, boosting rural development beyond sustainable agriculture. It is an investment with high returns. Improving rural productivity also requires significant investments in basic infrastructure, including transportation, rural energy, and irrigation. There will be little progress without such foundational investments.

Fostering entrepreneurship and facilitating private sector development must be highest on the agenda to promote the autonomy and support needed to translate opportunity into prosperity. This has to be seen as an investment in itself, with carefully tailored incentives and risk-sharing approaches supported by government.

Toward a New Regional Economic Vision

Contemporary history informs us that the main explanation for the success of the industrialized countries lies in their ability to learn how to improve performance in a diversity of social, economic, and political fields. In other words, the key to their success was their focus on practical knowledge and the associated improvements in skills needed to solve problems. They put a premium on learning based on historical experiences.[29]

One of the most reassuring aspects of a learner's strategy is that every generation receives a legacy of knowledge that it can harness for its own use. Every generation blends the new and the old and thereby charts its own developmental path, making debates about innovation and tradition irrelevant. Furthermore, discussions on the impact of intellectual property rights take on a new meaning if one considers the fact that the further away you are from the frontier of research, the larger is your legacy of technical knowledge. The challenge, therefore, is for Africa to think of research in adaptive terms, rather than simply focusing on how to reach parity with the technological front-runners. Understanding the factors that help countries to harness available knowledge is critical to economic transformation.

The advancement of information technology and its rapid diffusion in recent years could not have happened without basic telecommunication infrastructure. In addition, electronic information systems, which rely on telecommunications infrastructure, account for a substantial proportion of production and distribution activities in the secondary and tertiary sectors of the economy. It should also be noted that the poor state of Africa's telecommunications infrastructure has hindered the capacity of the region to make use of advances in fields such as geographical information sciences in sustainable development.

The emphasis on knowledge is guided by the view that economic transformation is a process of continuous improvement in productive activities. In other words, government policy should be aimed at enhancing performance, starting with critical fields such as agriculture, while recognizing interdisciplinary linkages.

This type of improvement indicates a society's capacity to adapt to change through learning. It is through continuous improvement that nations transform their economies and achieve higher levels of performance. Using this framework, with government functioning as a facilitator for economic learning, agribusiness enterprises will become the locus of learning, and knowledge will be the currency of change.

Some African countries already possess the key institutional components they need to become players in the knowledge economy. The emphasis, therefore, should be on realigning the existing structures, creating necessary new ones where they do not exist, and promoting interactions between key players in the economy. More specifically, the separation between government, industry, and academia stands out as one of the main sources of inertia and waste in Africa's knowledge-based institutions.[30] The challenge is not simply creating institutions, but creating systems of innovation in which emphasis is placed on economic learning through interactions between actors in the society.

A key role of Africa's RECs is to provide the regional framework for all stakeholders to act in a coordinated manner, share best practices, encourage peer review of achievements and setbacks by key players, and pool resources for the greater good of the region and Africa at large. The policy organs of the RECs, including the presidents and sectoral ministers, provide appropriate frameworks for the public and private sector to formulate innovative policies; given the multidisciplinary and multisectoral nature of the initiatives, the higher policy organs at the level of heads of state and government, and at the level of joint ministerial meetings, provide a unique role for the RECs as vehicles for promoting regional collaboration and for the elaboration and implementation of key policy initiatives.

Africa has visions for socioeconomic development at the national, regional, and continental level. Science, technology, engineering, and innovation are critical pillars of any socioeconomic development vision in our time. At the three levels, the visions do not coherently interact because the continental policies are not necessarily coordinated with the policies that the member states adopt and implement in the context of the RECs, and national policymaking is at times totally divorced from the regional and continental processes and frameworks.

However, there are case studies of how some RECs have tried to address this dilemma, which could constitute best practices for the implementation of regional policies at the national level and for the elaboration of regional policies on the basis of practical realities in the member states. In the East African Community (EAC), each member state has agreed to establish a dedicated full-scale ministry responsible for EAC affairs. This means that EAC affairs are organically integrated into the national government structure of the member states. There is need for a coherent approach to the formulation and implementation of regional policies at the national level, drawing on the collective wisdom and clout that RECs provide in tackling key national and regional challenges, particularly those related to the rapid socioeconomic transformation of Africa.

Agriculture and smart governance

One of the most important aspects of Africa's agricultural transformation is the role of leadership. Much of the attention on leadership has tended to focus on the achievements of presidents and prime ministers as such. As the case of Nigeria shows, leadership is primarily an institutional response to a challenge. Most of Africa's governance structures were created at a time when championing economic dynamism was not a priority. These structures have persisted despite the changed conditions. The challenge facing Africa leaders is therefore reforming government structures to make them more responsive to economic needs. There arc of course immediate challenges such as addressing corruption, improving transparency, and liberalizing markets. These efforts cannot be sustained without complementary institutional adjustments, some of which include the creation of new state organs.

Probably the most pressing need among African countries is creating new institutions that help leaders to make decisions based on the best available scientific and technical knowledge. This is particularly important for agriculture given its knowledge-intensive character. Systematic knowledge for decisionmaking is needed for every part of the agricultural value chain. Much of this knowledge exists in most countries. The knowledge, however, is not readily accessible to high-level leaders because of the absence of specialized science and technology advisory bodies. As of early 2015 no African president had a dedicated office of the chief science and innovation advisor. Nigeria's State Government had a Chief Scientist appointed in response to the Ebola crisis. African presidents or prime ministers need such officers. They can initially focus on agricultural transformation as part of a larger agenda of fostering enlightened economic governance.

The lessons learned through this initial phase will come in handy when addressing other economic transformation challenges such as health, education, industry, and services.

It is clear that Africa cannot muddle through its development agenda without systematic science and innovation advice. Doing so would be tantamount to going to battle without effective military intelligence. Defeat is almost guaranteed for those who fail to create institutions for enlightened leadership. Agriculture is Africa's best starting point for improving overall governance.

NOTES

Introduction

1. African Union, "Concept Note." Assembly of the Union, Twenty-second Ordinary Session, Addis Ababa, Ethiopia, January 30–31, 2014.

2. New Partnership for African Development (NEPAD), *Agriculture in Africa: Transformation and Outlook* (Johannesburg, South Africa: NEPAD, November 2013).

3. President Kikwete of Tanzania helped launch the first edition of this book and was subsequently instrumental in leading the advocacy for the creation of Grow Africa.

4. B. wa Mutharika, "Feeding Africa Through New Technologies: Let Us Act Now" (Acceptance Speech on Election of the Chairman of the Assembly of the African Union, Addis Ababa, January 31, 2010). As wa Mutharika explained: "I firmly believe that if we could agree that food security at the Africa level is a priority, then other priorities such as climate change, ICT, transport and infrastructure development would also become a necessity to enhance flow of information, movement of people, goods and services including the production and supply of agricultural inputs within and among nations, regions and the continent at large. I therefore propose that we consider investing in the construction of infrastructure to support food security. We need to build food storage facilities, new roads, railways, airlines, shipping industries as well as develop inter-state networks to ensure that we can move food surplus to deficit areas more efficiently and more cheaply."

5. C. P. Reij and E. M. Smaling, "Analyzing Successes in Agriculture and Land Management in Sub-Saharan Africa: Is Macro-Level Gloom Obscuring Micro-Level Change?" *Land Use Policy* 25, no. 3 (2008): 410–420.

6. Discussions on the role of innovation in development often ignore the role of engineering in development. For more details, see C. Juma, "Redesigning African Economies: The Role of Engineering in International Development" (Hinton Lecture, Royal Academy of Engineering, London, 2006); P. Guthrie, C. Juma, and H. Sillem, eds., *Engineering Change: Towards a Sustainable Future in the Developing World* (London: Royal Academy of Engineering, 2008).

7. H. T. Vesala and K. M. Vesala, "Entrepreneurs and Producers: Identities of Finnish Farmers in 2001 and 2006," *Journal of Rural Studies* 26, no. 1 (2010): 21–30.

8. The African Union (AU) recognizes the following RECs as the continent's economic integration building blocks: Community of Sahel Sahara States (CEN-SAD); Arab Maghreb Union (AMU); Economic Community of Central African States (ECCAS); Common Market of Eastern and Southern Africa (COMESA); Southern African Development Community (SADC); Intergovernmental Authority for Development (IGAD); Economic Community of West African States (ECOWAS); the East African Community (EAC).

9. Royal Society of London, *Reaping the Benefits: Science and the Sustainable Intensification of Global Agriculture* (London: Royal Society of London, 2009).

10. E. Kraemer-Mbula and W. Wamae, eds., *Innovation and the Development Agenda* (Paris: Organisation for Economic Co-operation and Development, 2010).

Chapter 1

1. This section is based on M. Fregene, personal communication with author, December 2014.

2. U. Lele, "Structural Adjustment, Agricultural Development and the Poor: Some Lessons from the Malawian Experience," *World Development* 18, no. 9 (1990): 1207–1219.

3. D. E. Sahn and J. Arulpragasam. "The Stagnation of Smallholder Agriculture in Malawi: A Decade of Structural Adjustment," *Food Policy* 16, no. 3 (1991): 219–234.

4. V. Quinn, "A History of the Politics of Food and Nutrition in Malawi: The Context of Food and Nutritional Surveillance," *Food Policy* 19, no. 3 (1994): 255–271.
5. G. Denning et al., "Input Subsidies to Improve Smallholder Maize Productivity in Malawi: Toward an African Green Revolution," *PLoS Biology* 7, no. 1 (2009): e1000023.
6. B. wa Mutharika, "Capitalizing on Opportunity" (paper presented at the World Economic Forum on Africa, Cape Town, South Africa, June 4, 2008).
7. B. Chinsinga, *Reclaiming Policy Space: Lessons from Malawi's Fertilizer Subsidy Programme* (London: Future Agricultures, 2007).
8. F. Ellis, *Fertiliser Subsidies and Social Cash Transfers: Complementary or Competing Instruments for Reducing Vulnerability to Hunger?* (Johannesburg, South Africa: Regional Hunger and Vulnerability Programme, 2009).
9. H. Ndilowe, personal communication, Embassy of Malawi, Washington, DC, 2009.
10. G. Denning et al. "Input Subsidies to Improve Smallholder Maize Productivity in Malawi: Toward an African Green Revolution," *PLoS Biology* 7, no. 1 (2009): e1000023.
11. D. C. Chibonga, personal communication, National Smallholder Farmers' Association, Malawi, 2009.
12. J. A. Schumpeter, *Capitalism, Socialism and Democracy* (New York: HarperCollins, 1942; reprint, with an introduction by Thomas K. McCraw, 2008), 82.
13. Calestous Juma, "Innovation and Development," 3–18.
14. C. P. Timmer, "Agriculture and Economic Development," *Handbook of Agricultural Economics* 2 (2002): 1487–1546.
15. P. Pingali, "Agricultural Growth and Economic Development: A View through the Globalization Lens," *Agricultural Economics* 37, no. 1 (2007): 1–12.
16. D. Byerlee, A. de Janvy, and E. Sadoulet, "Agriculture for Development: Toward a New Paradigm," *Annual Review of Resource Economics* 1, no.1 (2009): 15–35.
17. P. Pingali, "Agriculture Renaissance: Making 'Agriculture for Development' Work in the 21st Century," *Handbook of Agricultural Economics* 4 (2010): 3867–3894.
18. M. Tiffen, "Transition in Sub-Saharan Africa: Agriculture, Urbanization and Income Growth," *World Development* 31, no. 8 (2003): 1343–1366.

19. P. Gollin, "Agricultural Productivity and Economic Growth," *Handbook of Agricultural Economics* 4 (2010): 3825–3866.
20. For a comprehensive review of food-related debates, see R. Paarlberg, *Food Politics: What Everyone Needs to Know* (New York: Oxford University Press, 2010).
21. R. Evenson and D. Gollin, "Assessing the Impact of the Green Revolution, 1960–2000," *Science* 300, no. 5620 (2003): 758–762.
22. R. Evenson and D. Gollin, "Assessing the Impact of the Green Revolution, 1960–2000," *Science* 300, no. 5620 (2003): 758–762.
23. This is clearly articulated in InterAcademy Council, *Realizing the Promise and Potential of African Agriculture* (Amsterdam: InterAcademy Council, 2004).
24. P. Hazell, "An Assessment of the Impact of Agricultural Research in South Asia since the Green Revolution," *Handbook of Agricultural Economics* 4 (2010): 3469–3530.
25. *African Economic Outlook 2013: Structural Transformation and Natural Resources* (published 2014), 24–25; and Schaffnit-Chatterjee, Claire. 2014. "Agricultural Value Chains in Sub-Saharan Africa: From a Development Challenge to a Business Opportunity," *Deutsche Bank Research* (Frankfurt: Deutsche Bank AG, 2014).
26. World Bank, *World Development Report, 2008* (Washington, DC: World Bank, 2008).
27. L. Christiaesen, L. Demery, and J. Kuhl, "The (Evolving) Role of Agriculture in Poverty Reduction—An Empirical Perspective," *Journal of Development Economics* 96, no. 2 (2011): 250.
28. FAO, "Contribution of Agricultural Growth to Reduction of Poverty, Hunger and Malnutrition," in *State of Food Insecurity in the World 2012* (Rome: FAO, 2012), 28.
29. *African Economic Outlook 2013: Structural Transformation and Natural Resources,* 76.
30. World Bank, *World Development Report, 2008.* Washington, DC: World Bank, 2008, 40–41.
31. World Bank, *World Development Report, 2008.* Washington, DC: World Bank, 2008, 53.
32. World Bank, *World Development Report, 2008.* Washington, DC: World Bank, 2008, 27.
33. L. You et al., "What is the Irrigation Potential for Africa? A Combined Biophysical and Socioeconomic Approach," IFPRI Discussion Paper #00,993 (Washington, DC: International Food Policy Research Institute, 2010).

34. Organisation for Economic Co-operation and Development, "Agriculture, Food Security and Rural Development for Growth and Poverty Reduction: China's Agricultural Transformation— Lessons for Africa and its Development Partners." Summary of discussions by the China–DAC Study Group, Bamako, Mali, April 27–28, 2010.

35. United Nations Economic Commission for Africa and African Union, *Economic Report on Africa 2009: Developing African Agriculture Through Regional Value Chains* (Addis Ababa: United Nations Economic Commission for Africa and African Union, 2009).

36. United Nations Economic Commission for Africa and African Union, *Economic Report on Africa 2009: Developing African Agriculture Through Regional Value Chains* (Addis Ababa: United Nations Economic Commission for Africa and African Union, 2009), 117.

37. United Nations Economic Commission for Africa (UNECA), *Economic Report on Africa 2009: Developing African Agriculture Through Regional Value Chains* (Addis Ababa: UNECA, 2009).

38. United Nations Economic Commission for Africa, *Economic Report on Africa 2009: Developing African Agriculture Through Regional Value Chains* (Addis Ababa: UNECA, 2009), 117.

39. World Bank, *World Development Report 2008: Agriculture for Development* (Washington, DC: World Bank, 2007).

40. K. Gwilliam, V. Foster, R. Archondo-Callao, C. Briceño-Garmendia, A. Nogales, and K. Sethi, "Africa Infrastructure Country Diagnostic: Roads in Sub-Saharan Africa." Background Paper #14 (Washington, DC: World Bank et al., 2008); and World Bank, *World Development Report 2008: Agriculture for Development* (Washington, DC: World Bank, 2007).

41. World Bank, "Human Capital for Agriculture in Africa," *Science, Technology and Skills for Africa's Development* (Washington, DC: World Bank, March 2014).

42. World Bank, "Human Capital for Agriculture in Africa," *Science, Technology and Skills for Africa's Development* (Washington, DC: World Bank, March 2014), 2.

43. United Nations Economic Commission for Africa and African Union, *Economic Report on Africa 2009: Developing African Agriculture Through Regional Value Chains* (Addis Ababa: United Nations Economic Commission for Africa and African Union, 2009), 123, 125.

44. United Nations Economic Commission for Africa and African Union, *Economic Report on Africa 2009: Developing African Agriculture Through Regional Value Chains* (Addis Ababa: United Nations Economic Commission for Africa and African Union, 2009), 125–126.

45. R. Paarlberg, *Starved for Science: How Biotechnology Is Being Kept Out of Africa* (Cambridge, MA: Harvard University Press, 2008), 81.

46. United Nations Economic Commission for Africa and African Union, *Economic Report on Africa 2009: Developing African Agriculture Through Regional Value Chains* (Addis Ababa: United Nations Economic Commission for Africa and African Union, 2009), 126.

47. United Nations Economic Commission for Africa and African Union, *Economic Report on Africa 2009: Developing African Agriculture Through Regional Value Chains* (Addis Ababa: United Nations Economic Commission for Africa and African Union, 2009), 126–127.

48. AGRA, *Africa Agriculture Status Report 2014*, 142.

49. S. Tonassi, "Irrigating Africa" (Washington, DC: IFPRI, July 2010), http://www.ifpri.org/blog/irrigating-africa.

50. United Nations Economic Commission for Africa and African Union, *Economic Report on Africa 2009: Developing African Agriculture Through Regional Value Chains* (Addis Ababa: United Nations Economic Commission for Africa and African Union, 2009), 129.

51. African Development Bank, *Annual Development Effectiveness Report 2013: Towards Sustainable Growth for Africa* (Tunisia: African Development Bank Group, June 2013).

52. Annual Development Effectiveness Review 2013.

53. African Development Bank, *Annual Development Effectiveness Report 2013: Towards Sustainable Growth for Africa* (Tunisia: African Development Bank Group, June 2013), 26.

54. Quoted in World Bank, *World Development Report 2008: Agriculture for Development*. Washington, D.C.: World Bank, 2007: 229.

55. African Center for Economic Transformation, *Growth with Depth: The 2014 African Transformation Report* (Accra, Ghana: African Center for Economic Transformation, 2014).

56. Africa Progress Panel, *Grain, Fish, Money: Financing Africa's Green and Blue Revolutions* (Geneva: Africa Progress Panel 2014), 34.

57. Calestous Juma, "Growing the Nutritional Revolution: A Plea for Niche Crops," *Nestlé Foundation Report 2013* (Lausanne, Switzerland: Nestlé Foundation, May 2014).

58. African Center for Economic Transformation, *Growth with Depth: The 2014 African Transformation Report* (Accra, Ghana: African Center for Economic Transformation, 2014).

59. IFPRI, "Fish Farms to Produce Nearly Two Thirds of Global Food Fish Supply by 2030, Report Shows," February 3, 2014, http://www.ifpri.org/pressrelease/fish-farms-produce-nearly-two-thirds-global-food-fish-supply-2030-report-shows.

60. Africa Progress Panel, *Grain, Fish, Money: Financing Africa's Green and Blue Revolutions* (Geneva, Switzerland: Africa Progress Panel, 2014), 82.

61. AGRA, *Africa Agriculture Status Report 2014*, 66.

62. Nazaire Houssou, Xinshen Diao, and Shashi Kolavalli, "Can The Private Sector Lead Agricultural Mechanization in Ghana?" IFPRI Policy Note #4 (Washington, DC: IFPRI, August 2014).

63. Nazaire Houssou, Xinshen Diao, and Shashi Kolavalli, "Can the Private Sector Lead Agricultural Mechanization in Ghana?" IFPRI Policy Note #4 (Washington, DC: IFPRI, August 2014).

64. R. Roxburgh et al., *Lions on the Move: The Progress and Potential of African Economies* (Washington, DC: McKinsey Global Institute, 2010), 7–8.

65. Ousmane Badiane, Tsitsi Makombe, and Godfrey Bahiigwa, *Promoting Agricultural Trade to Enhance Resilience in Africa* (Washington, DC: IFPRI 2013).

66. Office of the Press Secretary, "Fact Sheet: G-8 Action on Food Security and Nutrition," White House, May 18, 2012, http://www.whitehouse.gov/the-press-office/2012/05/18/fact-sheet-g-8-action-food-security-and-nutrition.

67. See Grow Africa Secretariat, *Investing in the Future of African Agriculture: 1st Annual Report on Private-Sector Investment in Support of Country-Led Transformations in African Agriculture* (Geneva: World Economic Forum, 2013), 2, 11.

68. For much more detail on the specific investments being made, see Grow Africa Secretariat, *Agricultural Partnerships Take Root across Africa*, Annual Report, May 2014.

69. See, for example, George Monbiot, "Africa Let Us Help—Just like in 1884," *Guardian*, June 10, 2013, http://www.theguardian.com/commentisfree/2013/jun/10/african-hunger-help-g8-grab?guni=Article:in%20body%20link.

70. Grow Africa Secretariat, *Investing in the Future of African Agriculture*, 11. See also David Hong, "Debunking Myths: Five Things You (Probably) Didn't Know about the New Alliance and Food Security," ONE.org, December 10, 2012, http://www.one.org/us/2012/12/10/debunking-myths-five-things-you-probably-didnt-know-about-the-new-alliance-and-food-security/; and David Hong, "New Alliance for Food Security and Nutrition: Part 2," ONE.org, n.d., http://www.one.org/us/policy/new-alliance-for-food-security-and-nutrition-part-2/.

71. Molly Kinder and Nachilala Nkombo, "New Alliance as Imperialism in Another Guise? Monbiot Should Know Better," Poverty Matters blog, *Guardian*, June 12, 2013.

72. P. D. Williams and J. Haacke, "Security Culture, Transnational Challenges and the Economic Community of West African States," *Journal of Contemporary African Studies* 26, no. 2 (2008): 119–123.

73. OECD, *African Economic Outlook 2010: Public Resource Mobilization and Aid* (Paris: Organisation for Economic Cooperation and Development, 2010).

74. EAC, *Agriculture and Rural Development Strategy for the East African Community* (Arusha, Tanzania: East African Community, 2006).

75. EAC, *Agriculture and Rural Development Strategy for the East African Community* (Arusha, Tanzania: East African Community, 2006).

76. T. Persson and G. Tabellini, "The Growth Effect of Democracy: Is It Heterogeneous and How Can It Be Estimated?" In *Institutions and Economic Performance*, ed. E. Helpman (Cambridge, MA: Harvard University Press, 2008).

77. Africa Progress Panel, *Grain, Fish, Money: Financing Africa's Green and Blue Revolutions* (Geneva: Africa Progress Panel, 2014), 75.

78. Africa Progress Panel, *Grain, Fish, Money: Financing Africa's Green and Blue Revolutions* (Geneva: Africa Progress Panel, 2014).

79. Alliance for a Green Revolution (AGRA), *Africa Agriculture Status Report 2014: Climate Change and Smallholder Agriculture in sub-Saharan Africa* (Nairobi, Kenya: AGRA, 2014).

Notes 275

80. L. E. Fulginiti, "What Comes First, Agricultural Growth or Democracy?" *Agricultural Economics* 41, no. 1 (2010): 15–24.
81. R. Nelson, "What Enables Rapid Economic Progress: What Are the Needed Institutions?" *Research Policy* 37, no. 1 (2008): 1.
82. D. Restuccia, D. T. Yang, and X. Zhu, "Agriculture and Aggregate Productivity," *Journal of Monetary Economics* 55, no. 2 (2008): 234–250.

Chapter 2

1. P. H. Diamandis and S. Kotler, *Abundance: The Future is Better Than You Think* (New York: Free Press 2012), 34.
2. W. B. Arthur, *The Nature of Technology, What It Is and How it Evolves* (New York: Free Press, 2009).
3. P. H. Diamandis and S. Kotler, *Abundance: The Future is Better Than You Think* (New York: Free Press, 2012).
4. E. Brynjolfsson and A. McAfee, *The Second Machine Age: Work, Progress, and Prosperity in a Time of Brilliant Technologies* (New York; W.W. Norton).
5. J. Schwerin and C. Werker, "Learning Innovation Policy Based on Historical Experience," *Structural Change and Economic Dynamics* 14, no. 4 (2004): 385–404.
6. C. Freeman and F. Lou ç ã, *As Time Goes By: From the Industrial Revolution to the Information Revolution* (Oxford: Oxford University Press, 2001); J. Fegerberg et al., eds., *The Oxford Handbook of Innovation* (Oxford: Oxford University Press, 2005).
7. T. Ridley, Y.-C. Lee, and C. Juma, "Infrastructure, Innovation and Development," *International Journal of Technology and Globalisation* 2, no. 3 (2006): 268–278; D. Rouach and D. Saperstein, "Alstom Technology Transfer Experience: The Case of the Korean Train Express (KTX)," *International Journal of Technology Transfer and Commercialisation* 3, no. 3 (2004): 308–323.
8. B. Oyelaran-Oyeyinka and K. Lal, "Learning New Technologies by Small and Medium Enterprises in Developing Countries," *Techno-vation* 26, no. 2 (2006): 220–231.
9. See, for example, D. J. Isenberg, "How to Start an Entreprenaurial Revolution," *Harvard Business Review*, 2010.
10. C. Juma, "Redesigning African Economies: The Role of Engineering in International Development" (Hinton Lecture, Royal Academy of Engineering, London, 2006).

11. D. King, "Governing Technology and Growth," in *Going for Growth: Science, Technology and Innovation in Africa*, ed. C. Juma (London: Smith Institute, 2005), 112–124.

12. C. Juma, "Reinventing African Economies: Technological Innovation and the Sustainability Transition" (paper presented at the 6th John Pesek Colloquium on Sustainable Agriculture, Iowa State University, Ames, Iowa, 2006).

13. A. D. Alene, "Productivity Growth and the Effects of R&D in African Agriculture," *Agricultural Economics* 41, nos. 3–4 (2010): 223–238.

14. W. A. Masters, "Paying for Prosperity: How and Why to Invest in Agricultural R&D for Development in Africa," *Journal of International Affairs* 58, no. 2 (2005): 35–64.

15. H. Jikun and S. Rozelle, "Technological Change: Rediscovering the Engine of Productivity Growth in China's Rural Economy," *Journal of Development Economics* 49, no. 2 (1996): 337–369.

16. S. Fan and J. Brzeska, "Production, Productivity, and Public Investment in East African Agriculture," in *Handbook of Agricultural Economics*, Vol. 4 (Oxford: Elsevier, 2010), 3403.

17. S. Fan and J. Brzeska, "Production, Productivity, and Public Investment in East African Agriculture," in *Handbook of Agricultural Economics*, Vol. 4 (Oxford: Elsevier, 2010), 3403.

18. G. Conway and J. Waage, *Science and Innovation for Development* (London: UK Collaborative on Development Sciences, 2010), 37.

19. This section is based on S. Menker, personal communication with author, New York, December 2014.

20. International Telecommunication Union, *Measuring the Information in Society Report 2014*, Geneva: International Telecommunication Union, http://www.itu.int/en/ITU-D/ Statistics/Documents/publications/mis2014/MIS2014_without_ Annex_4.pdf.

21. International Telocommunication Union, *The World in 2014 ICT Facts and Figures*, Geneva: International Telecommunication Union, 2014, http://www.itu.int/en/ITU-D/Statistics/ Documents/facts/ICTFactsFigures2014-e.pdf.

22. J. C. Acker and I. M. Mbiti, *Mobile Phones and Economic Development in Africa* (Washington, DC: Center for Global Development, 2010).

23. G. Conway and J. Waage, *Science and Innovation for Development* (London: UK Collaborative on Development Sciences, 2010), 37.

24. G. Conway and J. Waage, *Science and Innovation for Development* (London: UK Collaborative on Development Sciences, 2010), 37.
25. International Telocommunication Union, *The World in 2014 ICT Facts and Figures*, Geneva: International Telecommunication Union, 2014, http://www.itu.int/en/ITU-D/Statistics/Documents/facts/ICTFactsFigures2014-e.pdf.
26. W. Jack and T. Suri, *Mobile Money: The Economics of M-PESA* (Cambridge, MA: Sloan School, Massachusetts Institute of Technology, 2009).
27. I. Mas, "The Economics of Branchless Banking," *Journal of Monetary Economics* 4, no. 2 (2009): 57–76.
28. M. L. Rilwani and I. A. Ikhuoria, "Precision Farming with Geoinformatics: A New Paradigm for Agricultural Production in a Developing Country," *Transactions in GIS* 10, no. 2 (2006): 177–197.
29. P. M. B. Waswa and C. Juma, "Establishing a Space Sector for Sustainable Development in Kenya." *International Journal of Technology and Globalization* 6, no 1–2 (2012).
30. R. Chwala, personal communication, Survey Settlements and Land Records Department, State of Karnataka, 2009.
31. C. Anderson, "Agricultural Drones." *MIT Technology Review*, April. 23, 2014, http://www.technologyreview.com/featuredstory/526491/agricultural-drones/.
32. T. Fleischer and A. Grunwald, "Making Nanotechnology Developments Sustainable: A Role for Technology Assessment?" *Journal of Cleaner Production* 16, nos. 8–9 (2008): 889–898.
33. G. Conway and J. Waage, *Science and Innovation for Development* (London: UK Collaborative on Development Sciences, 2010), 53.
34. S. Louis and D. Alcorta, personal communication, Dais Analytic Corporation, Odessa, Florida 2010.
35. J. Perkins, *Geopolitics and the Green Revolution: Wheat, Genes, and the Cold War* (New York: Oxford University Press, 1997).
36. B. Bell, Jr., and C. Juma, "Technology Prospecting: Lessons from the Early History of the Chile Foundation," *International Journal of Technology and Globalisation* 3, nos. 2–3 (2007): 296–314.
37. K. Kastenhofer, "Do We Need a Specific Kind of Technoscience Assessment? Taking the Convergence of Science and Technology Seriously," *Poiesis Prax* 7, nos. 1–2 (2010): 37–54.
38. W. Russell et al., "Technology Assessment in Social Context: The Case for a New Framework for Assessing and Shaping

Technological Development," *Impact Assessment and Project Appraisal* 28, no. 2 (2010): 109–116.

39. L. Pellizoni, "Uncertainty and Participatory Democracy," *Environmental Values* 12, no. 2 (2003): 195–224.

40. A. Hall, "Embedding Research in Society: Development Assistance Options for Agricultural Innovation in a Global Knowledge Economy," *International Journal of Technology Management and Sustainable Development* 8, no. 3 (2008): 221–235.

Chapter 3

1. David Baulcombe, Jim Dunwell, Jonathan Jones, John Pickett, and Pere Puigdomenech, *GM Science Update: A Report to Council for Science and Technology* (London: UK Council for Science and Technology, n.d.), 28.

2. Of the three GE techniques—Clustered Regularly Interspaced Short Palindromic Repeats (CRISPRs), Zinc Finger Nucleases (ZFNs), and Transcription Activator-like Effector Nucleases (TALENs)—the CRISPR method is the most efficient, cost-effective, and easy to use. Although CRISPRs have drawbacks—they are not yet precise or accurate enough to be used in widespread fruit breeding—they present the best innovative solution to sustainable agriculture without genetic modification.

3. Chidananda Nagamangala Kanchiswamy, Daniel James Sargent, Riccardo Velasco, Massimo E. Maffei, and Mickael Malnoy, "Looking Forward to Genetically Edited Fruit Crops," *Trends in Biotechnology* (forthcoming).

4. For more on genetic editing, see "Genetically 'Edited' Fruit Could Soon Hit Supermarket Shelves," *Telegraph*, August 13, 2014; Chidananda Nagamangala Kanchiswamy, Daniel James Sargent, Riccardo Velasco, Massimo E. Maffei, and Mickael Malnoy, "Looking Forward to Genetically Edited Fruit Crops," *Trends in Biotechnology* (forthcoming); Nicholas J. Baltes and Daniel F. Voytas, "Enabling Plant Synthetic Biology Through Genome Engineering," *Trends in Biotechnology* (forthcoming); and David Baulcombe, Jim Dunwell, Jonathan Jones, John Pickett, and Pere Puigdomenech, *GM Science Update: A Report to Council for Science and Technology* (London: UK Council for Science and Technology, n.d.).

5. For more on this, see Keun Lee, Calestous Juma, and John Mathews, "Innovation Capabilities for Sustainable Development

in Africa," WIDER Working Paper 2014/062 (Helsinki Finland: World Institute for Development Economics, March 2014).

6. C. James, *Global Status of Commercialized Biotech/GM Crops: 2014*. ISAAA Brief No. 49. Ithaca, NY: International Service for the Acquisition of Agri-biotech Applications, 2014.

7. C. James, *Global Status of Commercialized Biotech/GM Crops: 2014*. ISAAA Brief No. 49. Ithaca, NY: International Service for the Acquisition of Agri-biotech Applications, 2014; and B. Choudhary and K. Gaur, Biotech Cotton in India, 2002 to 2014. Ithaca, NY: ISAAA, 2015.

8. C. James, *Global Status of Commercialized Biotech/GM Crops: 2014*. ISAAA Brief No. 49. Ithaca, NY: International Service for the Acquisition of Agri-biotech Applications, 2014.

9. R. Kaplinsky et al., "Below the Radar: What Does Innovation in Emerging Economies Have to Offer Other Low-income Countries?" *International Journal of Technology Management and Sustainable Development* 8, no. 3 (2009): 177–197. Indian entrepreneurs have devised ways of doing more with less based on the principles of affordability and sustainability: C. P. Prahalad and R. A. Mashelkar, "Innovation's Holy Grain," *Harvard Business Review,* July–August (2010): 1–10.

10. J. D. Glover and J. P. Reganold, "Perenial Grains: Food Security for the Future," *Issues in Science and Technology* 26, no. 2 (2010): 41–47.

11. C. James, *Global Status of Commercialized Biotech/GM Crops: 2013*. ISAAA Brief No. 46. Ithaca, NY: International Service for the Acquisition of Agri-biotech Applications, 2013.

12. J. Huang, C. Pray, and S. Rozelle, "Enhancing the crops to feed the poor." *Nature* 418 (2002): 678.

13. For a more detailed analysis of the benefits of biotechnology, see Calestous Juma and Katherine Gordon, "Transgenic Crops and Food Security," in Agnès Ricroch, Surinder Chopra, and Shelby J. Fleischer, eds., *Plant Biotechnology* (London: Springer, 2014).

14. D. Zilberman, S. E. Sexton, M. Marra, and J. Fernandez-Cornejo. "The Economic Impact of Genetically Engineered Crops." *Choices* 25, no. 2 (2010): 1–25.

15. A. M. Showalter et al., "A Primer for Using Transgenic Insecticidal Cotton in Developing Countries," *Journal of Insect Science* 9 (2009): 1–39.

16. D. Zilberman, H. Ameden, and M. Qaim, "The Impact of
    Agricultural Biotechnology on Yields, Risks, and Biodiversity in
    Low-Income Countries," *Journal of Development Studies* 43, no. 1
    (2007): 63–78.
17. C. E. Pray, L. Nagarajan, J. Huang, R. Hu, and B. Ramaswami,
    "The Impact of Bt Cotton and the Potential Impact of
    Biotechnology on Other Crops in China and India." In *Frontiers
    of Economics and Globalization*, eds. C. A. Carter, G. Moschini, and
    I. Sheldon, Vol. 10 (London: Emerald, 2011), Ch. 41; and J. Kathage,
    and M. Qaim, "Economic Impacts and Impact Dynamics of
    Bt (bacillus thuringiensis) Cotton in India." *Proceedings of the
    National Academy of Sciences* 109, no. 29 (2012): 11652–11656.
18. D. Zilberman, S. E. Sexton, M. Marra, and J. Fernandez-Cornejo.
    "The Economic Impact of Genetically Engineered Crops."
    Choices 25, no. 2 (2010): 1–25.
19. C. E. Pray, L. Nagarajan, J. Huang, R. Hu, and B. Ramaswami,
    B., "The Impact of Bt Cotton and the Potential Impact of
    Biotechnology on Other Crops in China and India." In *Frontiers
    of Economics and Globalization*, eds. C. A. Carter, G. Moschini, and
    I. Sheldon, Vol. 10 (London: Emerald, 2011), Ch. 41.
20. G. Brookes and P. Barfoot, "The Global Income and Production
    Effects of Genetically Modified (GM) Crops 1996–2011." *GM
    Crops and Food: Biotechnology in Agriculture and the Food Chain* 4,
    no. 1 (2013): 74–83, at p. 82.
21. For more, see C. Juma, P. Conceição, and S. Levine,
    "Biotechnology and food security," in S. Smyth, PWB Philips,
    and D. Castle, eds., *Handbook on Agriculture, Biotechnology, and
    Development.* Edward Elgar: Cheltenham, 2014; and C. Juma,
    "Preventing hunger: Biotechnology is key," *Nature* 479 (2011).
22. Christopher J. M. Whittey, Monty Jones, Alan Tollervey, and
    Tim Wheeler, "Africa and Asia Need a Rational Debate on GM
    Crops," *Nature* (May 2, 2013).
23. Hammadi, Saad. 2014. Bangladeshi farmers caught in row over
    $600,000 GM aubergine trial. *Guardian*. June 4. http://www.
    theguardian.com/environment/2014/jun/05/gm-crop-
    bangladesh-bt-brinjal.
24. Eugenio, Butelli et al. "Enrichment of Tomato Fruit with Health-
    Promoting Anthocyanins by Expression of Select Transcription
    Factors." *Nature Biotechnology* 26, no. 11 (2008): 1301–1308; and
    David Shukman, "Genetically-Modified Purple Tomatoes

Heading for Shops." *BBC*, January 14, 2014, http://www.bbc.
com/news/science-environment-25,885,756.

25. C. James, "Executive Summary," in *Global Status of
Commercialized Biotech/GM Crops: 2013*, ISAAA Brief No. 46
(Ithaca, NY: International Service for the Acquisition of Agri-
Biotech Applications, 2013).

26. Gilles-Eric Séralini et al. "Long Term Toxicity of a Roundup
Herbicide and a Roundup-Tolerant Genetically Modified Maize."
Food and Chemical Toxicology, 50, no. 11 (2012): 4221–4231.

27. Gemma Arjó et al. "Plurality of Opinion, Scientific Discourse
and Pseudoscience: An In Depth Analysis of the Séralini et al.
Study Claiming That Roundup™ Ready Corn or the Herbicide
Roundup™ Cause Cancer in Rats." *Transgenic Research*, 22, no. 2
(2013): 255–267.

28. A. E. Ricroch, J. B. Bergé, and M. Kuntz, "Evaluation of
Genetically Engineered Crops Using Transcriptomic, Proteomic,
and Metabolomics Profiling Techniques." *Plant Physiology* 155,
no. 4 (2011): 1752–1761.

29. A. Nicolia, A. Manzo, F. Veronesi, and D. Rosellini, "An Overview of
the Last 10 Years of Genetically Engineered Crop Safety Research."
Critical Reviews in Biotechnology 34, no. 1 (2013): 77–88, at p. 81.

30. G. Conway and J. Waage, *Science and Innovation for Development*
(London: UK Collaborative on Development Sciences, 2010), 54.

31. For a more detailed discussion of agricultural entrepreneurship,
see C. Juma and D. J. Spielman, "Farmers as Entrepreneurs:
Sources Of Agricultural Innovation in Africa," in *New Directions
for Smallholder Agriculture*, eds. P. Hazell and A. Rahman (New
York: Oxford University Press, 2014).

32. A. Nicolia, A. Manzo, F. Veronesi, and D. Rosellini, "An
Overview of the Last 10 Years of Genetically Engineered Crop
Safety Research." *Critical Reviews in Biotechnology* 34, no. 1 (2013):
77–88 [published early online 23 September], http://www.
geneticliteracyproject.org/wp/wp-content/uploads/2013/10/
Nicolia-20,131.pdf.

33. A. Nicolia, A. Manzo, F. Veronesi, and D. Rosellini, "An
Overview of the Last 10 Years of Genetically Engineered Crop
Safety Research." *Critical Reviews in Biotechnology* 34, no. 1 (2013):
77–88 [published early online 23 September], <http://www.
geneticliteracyproject.org/wp/wp-content/uploads/2013/10/
Nicolia-20,131.pdf.

34. R. Falkner, "Regulating Biotech Trade: The Cartagena Protocol on Biosafety," *International Affairs* 76, no. 2 (2000): 299–313.
35. L. R. Ghisleri et al., "Risk Analysis and GM Foods: Scientific Risk Assessment," *European Food and Feed Law Review* 4, no. 4 (2009): 235–250.
36. T. Bernauer, *Genes, Trade, and Regulation: The Seeds of Conflict in Food Biotechnology* (Princeton, NJ: Princeton University Press, 2003). Most of the studies on the risks of agricultural biotechnology tend to focus on unintended negative impacts. But evidence of unintended benefits is emerging. See, for example, W. D. Hutchison et al., "Areawide Suppression of European Corn Borer with Bt Maize Reaps Savings to Non-Bt Maize Growers," *Science* 330, no. 6001 (2010): 222–225.
37. A. A. Adenle, E. J. Morris, and G. Parayil, "Status of Development, Regulation, and Adoption of GM Agriculture in Africa: View and Positions of Stakeholder Groups," *Food Policy* 43 (2013): 159–166.
38. Christopher J. M. Whittey, Monty Jones, Alan Tollervey, and Tim Wheeler, "Africa and Asia Need a Rational Debate on GM Crops," *Nature* (May 2, 2013).
39. Christopher J. M. Whittey, Monty Jones, Alan Tollervey, and Tim Wheeler, "Africa and Asia Need a Rational Debate on GM Crops," *Nature* (May 2, 2013).
40. T. Bernauer, *Genes, Trade, and Regulation: The Seeds of Conflict in Food Biotechnology* (Princeton, NJ: Princeton University Press, 2003). Most of the studies on the risks of agricultural biotechnology tend to focus on unintended negative impacts. But evidence of unintended benefits is emerging. See, for example, W. D. Hutchison et al., "Areawide Suppression of European Corn Borer with Bt Maize Reaps Savings to Non-Bt Maize Growers," *Science* 330, no. 6001 (2010): 222–225.
41. R. Paarlberg, *Starved for Science: How Biotechnology Is Being Kept Out of Africa* (Cambridge, MA: Harvard University Press, 2008), 2.
42. J. O. Adeoti and A. A. Adekunle, "Awareness of and Attitudes Towards Biotechnology and GMOs in Southwest Nigeria: A Survey of People with Access to Information," *International Journal of Biotechnology* 9, no. 2 (2007): 209–230.
43. E. J. Morris, "The Cartagena Protocol: Implications for Regional Trade and Technology Development in Africa," *Development Policy Review* 26, no. 1 (2008): 29–57.

44. "Kenya Food Industry toward Genetically Modified Food," *Food Policy* 35, no. 4 (2010): 332–340.
45. D. L. Kleinman, A. J. Kinchy, and R. Autry, "Local Variations or Global Convergence in Agricultural Biotechnology Policy? A Comparative Analysis," *Science and Public Policy* 36, no. 5 (2009): 361–371.
46. J. Keeley, "Balancing Technological Innovation and Environmental Regulation: An Analysis of Chinese Agricultural Biotechnology Governance," *Environmental Politics* 15, no. 2 (2000): 293–309.
47. Chidananda Nagamangala Kanchiswamy, Daniel James Sargent, Riccardo Velasco, Massimo E. Maffei, and Mickael Malnoy, "Looking Forward to Genetically Edited Fruit Crops," *Trends in Biotechnology* (forthcoming).
48. Erika Check Haydne, "Is the $1,000 Human Genome for real?" *Nature* (January 15, 2014), http://www.nature.com/news/is-the-1-000-genome-for-real-1.14530.

Chapter 4

1. G. E. Glasson et al., "Sustainability Science Education in Africa: Negotiating Indigenous Ways of Living with Nature in the Third Space," *International Journal of Science Education* 32, no. 1 (2010): 125–141.
2. S. W. Omamo and J. K. Lynam, "Agricultural Science and Technology Policy in Africa," *Research Policy* 32, no. 9 (2003): 1681–1694.
3. J. Fagerberg, "Introduction: A Guide to the Literature," in *The Oxford Handbook of Innovation*, ed. J. Fagerberg, D. Mowery, and R. Nelson (Oxford: Oxford University Press, 2005), 1–26.
4. J. Sumberg, "Systems of Innovation Theory and the Changing Architecture of Agricultural Research in Africa," *Food Policy* 30, no. 1 (2005): 21–41.
5. A. Hall, W. Janssen, E. Pehu, and R. Rajalahti, *Enhancing Agricultural Innovation: How to Go Beyond Strengthening Research Systems* (Washington, DC: World Bank, 2006), xiv.
6. A. Hall, W. Janssen, E. Pehu, and R. Rajalahti, *Enhancing Agricultural Innovation: How to Go Beyond Strengthening Research Systems* (Washington, DC: World Bank, 2006), xiv.
7. D. J. Spielman et al., "Developing the Art and Science of Innovation Systems Enquiry: Alternative Tools and Methods, and Applications to Sub-Saharan African Agriculture," *Technology in Society* 31, no. 4 (2009): 399–405.

8. J. O. Adeoti and O. Olubamiwa, "Toward an Innovation System in the Traditional Sector: The Case of the Nigerian Cocoa Industry," *Science and Public Policy* 36, no. 1 (2009): 15–31.

9. Scott Tiffin and Martin Kunc, "Measuring the Roles Universities Play in Regional Innovation Systems: A Comparative Study Between Chilean and Canadian Natural Resource-Based Regions," *Science and Public Policy* 38, no. 1 (February 2011): 55–66.

10. Scott Tiffin and Martin Kunc, "Measuring the Roles Universities Play in Regional Innovation Systems: A Comparative Study Between Chilean and Canadian Natural Resource-Based Regions," *Science and Public Policy* 38, no. 1 (February 2011): 55–66.

11. J. O. Adeoti, S. O. Odekunle, and F. M. Adeyinka, *Tackling Innovation Deficit: An Analysis of University-Firm Interaction in Nigeria* (Ibadan, Nigeria: Evergreen, 2010).

12. C. J. Maguire, *Technical Skills for Export Crop Industries in Uganda and Ethiopia* in *Agricultural Innovation Systems: An Investment Sourcebook* (Washington DC: The World Bank, 2012).

13. J. Ssebuwufu, T. Ludwick, and M. Béland, *Strengthening University-Industry Linkages in Africa: A Study of Institutional Capacities and Gaps* (Accra: Association of African Universities, 2012).

14. For more details, see J. Adeoti and O. Olubamiwa, "Toward an Innovation System in the Traditional Sector: The Case of the Nigerian Cocoa Industry," *Science and Public Policy* 36, no. 1 (2009): 15–31.

15. M. Gagné et al., "Technology Cluster Evaluation and Growth Factors: Literature Review," *Research Evaluation* 19, no. 2 (2010): 82–90.

16. M. S. Gertler and T. Vinodrai, "Life Sciences and Regional Innovation: One Path or Many?" *European Planning Studies* 17, no. 2 (2009): 235–261.

17. E. Gálvez-Nogale, *Agro-Based Clusters in Developing Countries: Staying Competitive in a Globalized Economy*, Agricultural Management, Marketing and Finance Occasional Paper. (Rome: UN Food and Agriculture Organization, 2010).

18. M. E. Porter, "Clusters and the New Economics of Competition," *Harvard Business Review* 76, no. 6 (1998): 77–90.

19. Z. Yingming, *Analysis of Industrial Clusters in China* (Boca Raton, FL: CRC Press, 2010).

20. This case study draws heavily from G. Wu, T. Tu, and S. Gu, *Innovation System and Transformation of the Agricultural Sector in*

China, with the Case of Shouguang City (paper presented to the
Globelics Conference on Innovation and Development, Rio
de Janeiro, November 3–6, 2003); S. Gu, "The Emergence and
Development of the Vegetable Sector in China," *Industry and
Innovation* 26, nos. 4–5 (2009): 499–524.

21. D. Dalohoun, A. Hall, and P. Mele, "Entrepreneurship as
Driver of a 'Self-Organizing System of Innovation': The
Case of NERICA in Benin," *International Journal of Technology
Management and Sustainable Development* 8, no. 2 (2009): 87–101.

22. Elisa Giuliani and Valeria Arza, "What Drives the Formation
of 'Valuable' University–Industry Linkages? Insights from the
Wine Industry," *Research Policy* 39 (2009): 906–921.

23. E. Gálvez-Nogale, *Agro-Based Clusters in Developing Countries:
Staying Competitive in a Globalized Economy*, Agricultural
Management, Marketing and Finance Occasional Paper (Rome:
UN Food and Agriculture Organization, 2010).

24. E. Gálvez-Nogale, *Agro-Based Clusters in Developing Countries:
Staying Competitive in a Globalized Economy*, Agricultural
Management, Marketing and Finance Occasional Paper (Rome:
UN Food and Agriculture Organization, 2010).

25. Elisa Giuliani, Andrea Morrison, Carlo Pietrobelli, and Roberta
Rabellotti, "Why Do Researchers Collaborate with Industry?
An Analysis of the Wine Sector in Chile, South Africa, and
Italy," CREI Working Paper no. 1/2009 (Rome: Centro di Ricerca
Interdipartimentale di Economia delle Istituzioni, Universita
Degli Studi, 2009).

26. P. Maskell, "Towards a Knowledge-Based Theory of the
Geographical Cluster," *Industrial and Corporate Change* 10, no. 4
(2001): 921.

27. S. Breschi and F. Malerba, "The Geography of Innovation and
Economic Clustering: Some Introductory Notes," *Industrial and
Corporate Change* 10, no. 4 (2001): 817.

28. B. Orlove et al., "Indigenous Climate Change Knowledge in
Southern Uganda," *Climatic Change* 100, no. 2 (2010): 243–265.

29. J. F. Martin et al., "Traditional Ecological Knowledge (TEK):
Ideas, Inspiration, and Design for Ecological Engineering,"
Ecological Engineering 36, no. 7 (2010): 839–849.

30. G. Glasson et al., "Sustainability Science Education in Africa:
Negotiating Indigenous Ways of Living with Nature in the
Third Space," *International Journal of Science Education* 32, no. 1
(2010): 125–141.

31. E. Ostrom, *Understanding Institutional Diversity* (Princeton, NJ: Princeton University Press, 2005).
32. Z. Xiwei and Y. Xiangdong, "Science and Technology Policy Reform and Its Impact on China's National Innovation System," *Technology in Society* 29, no. 3 (2007): 317.
33. Z. Xiwei and Y. Xiangdong, "Science and Technology Policy Reform and Its Impact on China's National Innovation System," *Technology in Society* 29, no. 3 (2007): 321.
34. X. Liu and T. Zhi, "China Is Catching Up in Science and Innovation: The Experience of the Chinese Academy of Sciences," *Science and Public Policy* 37, no. 5 (2010): 331–42.
35. Z. Xiwei and Y. Xiangdong, "Science and Technology Policy Reform and Its Impact on China's National Innovation System," *Technology in Society* 29, no. 3 (2007): 322.
36. N. Rada and C. Valdes, *Policy, Technology, and Efficiency of Brazilian Agriculture*, US Department of Agriculture Economic Research Service. Economic Research Report Number 137, July 2012.
37. M. A. Lopes and P. B. Arcuri, *The Brazilian Agricultural Research for Development (ARD) System* (paper presented at the International Workshop on Fast Growing Economies' Role in Global Agricultural Research for Development [ARD], Beijing, China, February 8–10, 2010).
38. M. A. Lopes and P. B. Arcuri, *The Brazilian Agricultural Research for Development (ARD) System* (paper presented at the International Workshop on Fast Growing Economies' Role in Global Agricultural Research for Development [ARD], Beijing, China, February 8–10, 2010).
39. M. A. Lopes and P. B. Arcuri, *The Brazilian Agricultural Research for Development (ARD) System* (paper presented at the International Workshop on Fast Growing Economies' Role in Global Agricultural Research for Development [ARD], Beijing, China, February 8–10, 2010).

Chapter 5

1. E. B. Barrios, "Infrastructure and Rural Development: Household Perceptions on Rural Development," *Progress in Planning* 70, no. 1 (2008): 1–44.
2. P. R. Agenor, "A Theory of Infrastructure-Led Development," *Journal of Economic Dynamics and Control* 34, no. 5 (2010): 932–950.
3. R. G. Teruel and Y. Kuroda, "Public Infrastructure and Productivity Growth in Philippine Agriculture, 1974–2000,"

Journal of Asian Economics 16, no. 3 (2005): 555–576; P.-R. Agenor and B. Moreno-Dodson, *Public Infrastructure and Growth: New Channels and Policy Implications* (Washington, DC: World Bank, 2006).
4. S. Fan and X. Zhang, "Public Expenditure, Growth and Poverty Reduction in Rural Uganda," *Africa Development Review* 20, no. 3 (2008): 466–496.
5. Ghana Millennium Development Authority, "The Compact," http://mida.gov.gh/site/?page_id=184.
6. Mark Tran, "Rwanda Rail Project on Track to Bridge Africa's Economic Divide," *The Guardian*, September 30, 2013.
7. See "Amendment to the Millennium Challenge Compact Between the United States of America Acting Through the Millennium Challenge Corporation and the Government of the Republic of Mali," http://www.mcc.gov/documents/agreements/compact-mali-01-amendment.pdf.
8. S. Fan and C. Chan-Kang, *Road Development, Economic Growth, and Poverty Reduction in China* (Washington, DC: International Food Policy Research Institute, 2005).
9. S. Fan and C. Chan-Kang, *Road Development, Economic Growth, and Poverty Reduction in China* (Washington, DC: International Food Policy Research Institute, 2005), vi.
10. Itai Madamombe, "Energy Key to Africa's Prosperity: Challenges in West Africa's Quest for Electricity," *Africa Renewal* 18, no. 4 (January 2005), http://www.geni.org/globalenergy/library/media_coverage/africa-renewal/energy-key-to-africas-prosperity.shtml.
11. World Bank, "West Africa Power Pool APL 1," Report No.: AB1450, 2005, pp. 2–3, http://www-wds.worldbank.org/servlet/WDSContentServer/IW3P/IB/2005/05/02/000104615_20050503101 210/Rendered/PDF/WAPP1PID1Appraisal0Stage.pdf.
12. The rest of this section is based on K. N. Gratwick and A. Eberhard, "An Analysis of Independent Power Projects in Africa: Understanding Development and Investment Outcomes," *Development Policy Review* 26, no. 3 (2008): 309–338.
13. Tam Harbert, "Construction Begins on $7 Billion Power Africa Project," *IEEE Spectrum*, February 19, 2014.
14. Ashley Dean, "Solar-Powered Irrigation Systems Improve Diet and Income in Rural Sub-Saharan Africa, Stanford Study Finds," *Stanford Report*, January 6, 2010.
15. Jose Simas, Juan Morelli, and Hani El Sadani, "Egypt: Irrigation Innovations in the Nile Delta," in *Water in the Arab*

World: Management Perspectives and Innovations, eds. N. Vijay
Jagannathan, Ahmed Shawky Mohamed, Alexander Kremer
(Washington, DC: World Bank, 2009), 433–442.

16. A. Narayanamoorthy, "Economics of Drip Irrigation in Sugarcane
Cultivation: Case Study of a Farmer from Tamil Nadu," *Indian
Journal of Agricultural Economics* 60, no. 2 (2005): 235.

17. See A. Narayanamoorthy, "Economics of Drip Irrigation
in Sugarcane Cultivation: Case Study of a Farmer from
Tamil Nadu," *Indian Journal of Agricultural Economics* 60, no. 2
(2005): 235.

18. K. Annamalai and S. Rao, *ITC's e-Choupal and the Profitable Rural
Transformation* (Washington, DC: World Resource Institute
[WRI], Michigan Business School, and UNC Kenan Flagler
Business School,).

19. For more on e-Choupal, see K. Annamalai and S. Rao, *ITC's
e-Choupal and the Profitable Rural Transformation* (Washington,
DC: World Resource Institute [WRI], Michigan Business School,
and UNC Kenan Flagler Business School,).

20. "Mobile Communications to Revolutionize African Weather
Monitoring," Press Release, June 18, 2009, http://hugin.
info/1061/R/1323500/310555.pdf.

21. This section draws heavily from D. Rouach and D. Saperstein,
"Alstom Technology Transfer Experience: The Case of the
Korean Train Express (KTX)," *International Journal of Technology
Transfer and Commercialisation* 3, no. 3 (2004): 308–323.

22. T. E. Mutambara, "Regional Transport Challenges with the
South African Development Community and Their Implications
for Economic Integration and Development," *Journal of
Contemporary African Studies* 27, no. 4 (2009): 501–525.

23. E. Calitz and J. Fourie, "Infrastructure in South Africa: Who Is to
Finance and Who Is to Pay?" *Development Southern Africa* 27, no.
2 (2010): 177–191.

24. K. Putranto, D. Stewart, and G. Moore, "International Technology
Transfer and Distribution of Technology Capabilities: The Case
of Railway Development in Indonesia," *International Journal of
Technology Transfer and Commercialisation* 3, no. 3 (2003): 43–53.

25. R. Shah and R. Batley, "Private-Sector Investment in
Infrastructure: Rationale and Causality for Pro-poor Impacts,"
Development Policy Review 27, no. 4 (2009): 397–417.

Chapter 6

1. A. Pinkerton and K. Dodds, "Radio Geopolitics: Broadcasting, Listening and the Struggle for Acoustic Space," *Progress in Human Geography* 33, no. 1 (2009): 10–27.
2. N. T. Assié-Lumumba, *Empowerment of Women in Higher Education in Africa: The Role and Mission of Research* (Paris: United Nations Educational, Scientific and Cultural Organization, 2006).
3. FAO, *The State of Food and Agriculture 2010–2011: Women in Agriculture* (Rome: FAO 2011).
4. For more examples, see FAO, *The State of Food and Agriculture 2010–2011: Women in Agriculture* (Rome: FAO 2011), 34–38.
5. For more on URDT and its role in training leaders, see Peter Senge, "The Solutions in Our Midst," *Oxford Leadership Journal* 1, no. 4 (October 2010).
6. M. Slavik, "Changes and Trends in Secondary Agricultural Education in the Czech Republic," *International Journal of Educational Development* 24, no. 5 (2004): 539–545.
7. J. Ndjeunga et al., *Early Adoption of Modern Groundnut Varieties in West Africa* (Hyderabad, India: International Crops Research in Semi-Arid Tropics, 2008).
8. G. Feder, R. Murgai, and J. Quizon, "Sending Farmers Back to School: The Impact of Farmer Field Schools in Indonesia," *Review of Agricultural Economics* 26, no. 1 (2004): 45–62.
9. P. Woomer, M. Bokanga, and G. Odhiambo, "*Striga* Management and the African Farmer," *Outlook on Agriculture* 37, no. 4 (2008): 277–282.
10. G. McDowell, *Land-Grant Universities and Extension into the 21st Century: Renegotiating or Abandoning a Social Contract* (Ames: Iowa State University Press, 2001).
11. H. Etzkowitz, "The Evolution of the Entrepreneurial University," *International Journal of Technology and Globalisation* 1, no. 1 (2004): 64–77.
12. M. Almeida, "Innovation and Entrepreneurship in Brazilian Universities," *International Journal of Technology Management and Sustainable Development* 7, no. 1 (2008): 39–58.
13. W. J. Mitsch et al., "Tropical Wetlands for Climate Change Research, Water Quality Management and Conservation Education on a University Campus in Costa Rica," *Ecological Engineering* 34, no. 4 (2008): 276–288.

14. The details on EARTH University are derived from C. Juma, "Agricultural Innovation and Economic Growth in Africa: Renewing International Cooperation," *International Journal of Technology and Globalisation* 4, no. 3 (2008): 256–275, and from updates provided by the President's Office at EARTH University.
15. M. Miller, M. J. Mariola, and D. O. Hansen, "EARTH to Farmers: Extension and the Adoption of Environmental Technologies in Humid Tropics of Costa Rica," *Ecological Engineering* 34, no. 4 (2008): 349–357.
16. National Research Council, *Transforming Agricultural Education for a Changing World* (Washington, DC: National Academies Press, 2009).
17. Alliance for a Green Revolution in Africa (AGRA), *Africa agriculture status report: Climate change and smallholder agriculture in sub-Saharan Africa* (Nairobi: AGRA, 2014): 43.
18. A. deGrassi, "Envisioning Futures of African Agriculture: Representation, Power, and Socially Constituted Time," *Progress in Development Studies* 7, no. 2 (2007): 79–98.
19. D. E. Leigh and A. M. Gill, "How Well Do Community Colleges Respond to the Occupational Training Needs of Local Communities? Evidence from California," *New Directions for Community Colleges* 2009, no. 146 (2009): 95–102.
20. M. Carnoy and T. Luschei, "Skill Acquisition in 'High Tech' Export Agriculture: A Case Study of Lifelong Earning in Peru's Asparagus Industry," *Journal of Education and Work* 21, no. 1 (2008): 1–23.

Chapter 7

1. World Bank, *Growing Africa: Unlocking the Potential of Agribusiness* (Washington, DC: World Bank, 2013).
2. Derek Byerlee, Andres F. Garcia, Asa Giertz, Vincent Palmade, *Growing Africa: Unlocking the Potential of Agribusiness*, Vol. 1 (Washington, DC: World Bank, 2013): xv.
3. Byerlee et al., *Growing Africa: Unlocking the Potential of Agribusiness*, 16.
4. Byerlee et al., *Growing Africa: Unlocking the Potential of Agribusiness*, 10.
5. Byerlee et al., *Growing Africa: Unlocking the Potential of Agribusiness*, 9.

6. Byerlee et al., *Growing Africa: Unlocking the Potential of Agribusiness*, 11.

7. J. Lerner, *Boulevard of Broken Dreams: Why Public Efforts to Boost Entrepreneurship and Venture Capital Have Failed—and What to Do About It* (Princeton, NJ: Princeton University Press, 2009).

8. A. Zacharakis, D. A. Shepherd, and J. A. Coombs, "The Development of Venture-Capital-Backed Internet Companies: An Ecosystem Perspective," *Journal of Business Venturing* 18, no. 2 (2003): 217–231.

9. G. Avnimelech, A. Rosiello, and M. Teubal, "Evolutionary Interpretation of Venture Capital Policy in Israel, Germany, UK and Scotland," *Science and Public Policy* 37, no. 2 (2010): 101–112.

10. Y. Huang, *Capitalism with Chinese Characteristics: Entrepreneurship and the State* (New York: Cambridge University Press, 2008).

11. Y. Huang, *Capitalism with Chinese Characteristics: Entrepreneurship and the State* (New York: Cambridge University Press, 2008).

12. Z. Zhang, "Rural Industrialization in China: From Backyard Furnaces to Township and Village Enterprises," *East Asia* 17 no. 3 (1999): 61–87.

13. S. Rozelle, J. Huang, and L. Zhang, "Emerging Markets, Evolving Institutions, and the New Opportunities for Growth in China's Rural Economy," *China Economic Review* 13, no. 4 (2002): 345–353.

14. H. Chen and S. Rozelle, "Leaders, Managers, and the Organization of Township and Village Enterprises in China," *Journal of Development Economics* 60, no. 2 (1999): 529–557.

15. E. Mabaya, "Strengthening Africa's Private Seed Sector to Serve Smallholder Farmers." Editorial, seedquest.com. July 2006.

16. A. S. Langyintuo et al., "Challenges of the Maize Seed Industry in Eastern and Southern Africa: A Compelling Case for Private-Public Intervention to Promote Growth," *Food Policy* 35, no. 4 (2010): 323–331.

17. Mark Tran, "Uganda Seed Entrepreneur Calls Time on Hand-Held Hoe as a Tool for Farmers," *The Guardian*, June 17, 2013.

18. C. E. Pray and L. Nagarajan, *Pearl Millet and Sorghum Improvement in India* (Washington, DC: International Food Policy Research Institute, 2009).

19. C. E. Pray and L. Nagarajan, *Pearl Millet and Sorghum Improvement in India* (Washington, DC: International Food Policy Research Institute, 2009).

20. M. Blackie, "Output to Purpose Review: Seeds of Development Programme (SODP)" (paper commissioned by the UK Department for International Development, London, August 7, 2008): 3.

21. M. Blackie, "Output to Purpose Review: Seeds of Development Programme (SODP)" (paper commissioned by the UK Department for International Development, London, August 7, 2008): 3–4.

22. J. Wilkinson, "The Food Processing Industry, Globalization and Development Economics," *Journal of Agricultural and Development Economics* 1, no. 2 (2004): 184–201.

23. For more information, see A. Dannson et al., *Strengthening Farm-Agribusiness Linkages in Africa: Summary Results of Five Country Studies in Ghana, Nigeria, Kenya, Uganda and South Africa* (Rome: UN Food and Agriculture Organization, 2004), 40–41.

24. For more information, see A. Dannson et al., *Strengthening Farm-Agribusiness Linkages in Africa: Summary Results of Five Country Studies in Ghana, Nigeria, Kenya, Uganda and South Africa* (Rome: UN Food and Agriculture Organization, 2004), 30.

25. This section based on M. Fregene and L. Pereira, personal communication with author, December 2014.

26. FMANR, *Action Plan for a Cassava Transformation in Nigeria* (2012), Federal Ministry of Agriculture and Natural Resources, Nigeria.

27. FAOSTAT, Food and Agricultural Organization of the United Nations (2013), http://faostat.fao.org/.

28. FMANR, *Action Plan for a Cassava Transformation in Nigeria* (2012), Federal Ministry of Agriculture and Natural Resources, Nigeria.

29. Taiwo, K. A., Oladepo, W. O., Ilori, M. O., Akanbi, C. T. 2002. A study on the Nigerian food industry and the impact of technological changes on the small-scale food enterprises. Food Reviews International, 18(4):243–262.

30. L. S. O. Liverpool, G. B. Ayoola, R. O. Oyeleke, "Enhancing the Competitiveness of Agricultural Commodity Chains in Nigeria: Identifying Opportunities with Cassava, Rice and Maize using a Policy Analysis Matrix (PAM) Framework," Nigeria Strategy Support Program (NSSP) Background Paper NSSP 13 (2009), International Food Policy Research Institute, Abuja, Nigeria.

31. M. U. Ukwuru and S. E. Egbono, "Recent Development in Cassava-Based Products Research," *Journal of Food Research* 1, no. 1 (2013): 1–13.

32. FMANR, Action Plan for a Cassava Transformation in Nigeria
    (2012), Federal Ministry of Agriculture and Natural Resources,
    Nigeria.

33. FMANR, *Action Plan for a Cassava Transformation in Nigeria*
    (2012), Federal Ministry of Agriculture and Natural Resources,
    Nigeria.

34. Ministry of Agriculture, Nigeria, personal communication with
    author, December 2014.

35. A. Van Oppen, "Cassava, 'The Lazy Man's Food'? Indigenous
    Agricultural Innovation and Dietary Change in Northwestern
    Zambia (ca. 1650–1970)," *Food and Foodways* 5, no. 1 (1991): 15–38.

36. "Radio, Video Key to Agricultural Innovation in Africa." *African
    Agriculture* blog, June 22, 2009, http://www.africanagricultureblog.
    com/2009/06/radio-video-key-to-agricultural.html.

37. "Radio, Video Key to Agricultural Innovation in Africa." *African
    Agriculture* blog, June 22, 2009, http://www.africanagricultureblog.
    com/2009/06/radio-video-key-to-agricultural.html.

38. "Radio, Video Key to Agricultural Innovation in Africa." *African
    Agriculture* blog, June 22, 2009, http://www.africanagricultureblog.
    com/2009/06/radio-video-key-to-agricultural.html.

39. "Radio, Video Key to Agricultural Innovation in Africa."
    African Agriculture blog, June 22, 2009, http://www.
    africanagricultureblog.com/2009/06/radio-video-key-to-
    agricultural.html. See also P. Van Mele, J. Wanvoeke, and
    E. Zossou, "Enhancing Rural Learning, Linkages, and
    Institutions: The Rice Videos in Africa," *Development in Practice*
    20, no. 3 (2010): 414–421.

40. This section is based on Su Kahumbu Stephanou, personal
    communication with author, December 2014.

41. J. Kerlin, "A Comparative Analysis of the Global Emergence of
    Social Enterprise," *Voluntas: International Journal of Voluntary and
    Nonprofit Organizations* 21, no. 2 (2010): 162–179.

42. This section is based on S. Hanson and H. Paulson, personal
    communication with author, Bungoma, Kenya, One Acre Fund,
    December 2014.

43. This section is based on M. Mehta and S. Kernan, personal
    communication with author, Tomato Jos, Nigeria, December
    2014.

44. This section is based on A. Oppenheim, personal
    communication with author, Vital Capital, December 2014.

45. A. Gupta, personal communication, Society for Research and Initiatives for Sustainable Technologies and Institutions, Ahmedabad, India, 2010.
46. This section is based on D. Ragone, personal communication with author, Breadfruit Institute, Hawaii, December 2014.
47. Andrew M. P. Jones, Diane Ragone, Kamaui Aiona, W. Alex Lane, and Susan J. Murch, "Nutritional and Morphological Diversity of Breadfruit (*Artocarpus*, Moraceae): Identification of Elite Cultivars for Food Security," *Journal of Food Composition and Analysis* 24, no. 8 (2011): 1091–1102; and A. Jones, P. Maxwell, Ross Baker, Diane Ragone, and Susan J. Murch, "Identification of Pro-Vitamin A Carotenoid-Rich Cultivars of Breadfruit (*Artocarpus*, Moraceae)," *Journal of Food Composition and Analysis* 31, no. 1 (2012): 51–61.
48. Julia Flynn Silver, "'Food of the Future' Has One Hitch: It's All But Inedible," *Wall Street Journal*, November 1, 2011.
49. Julia Flynn Silver, "'Food of the Future' Has One Hitch: It's All But Inedible," *Wall Street Journal*, November 1, 2011.
50. Ying Liu, A. Maxwell, P. Jones, Susan J. Murch, and Diane Ragone, "Crop Productivity, Yield, and Seasonality of Breadfruit (*Artocarpus* spp.) Moraceae," *Fruits* 69 (2014): 345–361.
51. Diane Ragone, "Breadfruit: *Artocarpus altilis* (Parkinson) Fosberg. Promoting the Conservation and Use of Underutilized and Neglected Crops 10" (Rome, Italy: International Plant Genetic Resources Institute, 1997).
52. This section is based on K. Masha, personal communication with author, Babban Gona, Nigeria, December 2014.

Chapter 8

1. M. Farrell, "EU Policy Towards Other Regions: Policy Learning in the External Promotion of Regional Integration," *Journal of European Public Policy* 16, no. 8 (2009): 1165–1184.
2. C. Juma and I. Serageldin, *Freedom to Innovate: Biotechnology in Africa's Development* (Addis Ababa: African Union and New Partnership for Africa's Development, 2007), 20.
3. C. Juma, C. Gitta, and A. DiSenso, "Forging New Technological Alliances: The Role of South-South Cooperation," *Cooperation South Journal* (2005): 59–71.
4. Commission for Africa, *Our Common Interest: Report of the Commission for Africa* (London: Commission for Africa, 2005).

5. Y. H. Kim, "The Optimal Path of Regional Economic Integration between Asymmetric Countries in the North East Asia," *Journal of Policy Modeling* 27, no. 6 (2005): 673–687.

6. "23rd AU Summit: Agriculture and Food Security, critical priority for Africa," www.cta.int, July 7, 2014.

7. "2014 Year of Agriculture." 22nd AU Summit: Agriculture and Food Security, http://pages.au.int/caadpyoa.

8. African Union, "Malabo Declaration on Accelerated Agricultural Growth and Transformation for Shared Prosperity and Improved Livelihoods," tralac.org, July 11, 2014.

9. G. Jones and S. Corbridge, "The Continuing Debate about Urban Bias: The Thesis, Its Critics, Its Influence and Its Implications for Poverty-Reduction Strategies," *Progress in Development Studies* 10, no. 1 (2010): 1–18.

10. L. M. Póvoa and M. S. Rapini, "Technology Transfer from Universities and Public Research Institutes to Firms in Brazil: What Is Transferred and How the Transfer Is Carried Out," *Science and Public Policy* 37, no. 2 (2010): 147–159.

11. S. Pitroda, personal communication, New Delhi, National Innovation Council, 2010.

12. C. Juma and Y.-C. Lee, *Innovation: Applying Knowledge in Development* (London: Earthscan, 2005), 140–158.

13. A. Sindzingre, "Financing the Developmental State: Tax and Revenue Issues," *Development Policy Review* 25, no. 5 (2007): 615–632.

14. N. Bloom, R. Griffith, and J. Van Reenen, "Do R&D Tax Credits Work? Evidence from a Panel of Countries 1979–1997," *Journal of Public Economics* 85, no. 1 (2002): 1–31.

15. B. Aschhoff and W. Sofka, "Innovation on Demand—Can Public Procurement Drive Market Success of Innovation?" *Research Policy* 38, no. 8 (2009): 1235–1247.

16. McKinsey & Company, *"And the Winner Is . . . ": Capturing the Promise of Philanthropic Prizes.* McKinsey & Company, 2009, http://www. mckinsey.com/App_Media/Reports/SSO/And_the_winner_is.pdf.

17. W. A. Masters and B. Delbecq, *Accelerating Innovation with Prize Rewards* (Washington, DC: International Food Policy Research Institute, 2008).

18. T. Cason, W. Masters, and R. Sheremeta, "Entry into Winner-Take-All and Proportional-Prize Contests: An Experimental Study," *Journal of Public Economics* 94, nos. 9–10 (2010): 604–611.

19. Philip G. Pardey and P. Pingali, "Reassessing International Agricultural Research for Food and Agriculture" (report prepared for the Global Conference on Agricultural Research for Development [GCARD], Montpellier, France, March 28–31, 2010); J. M. Alston, M. A. Andersen, J. S. James, and P. G. Pardey, *Persistence Pays: U.S. Agricultural Productivity Growth and the Benefits from Public R&D Spending* (New York: Springer, 2010).
20. R. E. Just, J. M. Alston, and D. Zilberman, eds., *Regulating Agricultural Biotechnology: Economics and Policy* (New York: Springer-Verlag, 2006).
21. W. B. Arthur, *The Nature of Technology: What It Is and How It Evolves* (New York: Free Press, 2009).
22. D. Archibugi and C. Pietrobelli, "The Globalisation of Technology and Its Implications for Developing Countries," *Technological Forecasting and Social Change* 70, no. 9 (2003): 861–883.
23. Y. Chen and T. Puttitanum, "Intellectual Property Rights and Innovation in Developing Countries," *Journal of Development Economics* 78, no. 2 (2005): 474–493.
24. C. E. Pray and N. Anwar, "Supplying Crop Biotechnology to the Poor: Opportunities and Constraints," *Journal of Development Studies* 43, no. 1 (2007): 192–217.
25. S. Redding and P. Schott, "Distance, Skill Deepening and Development: Will Peripheral Countries Ever Get Rich?" *Journal of Development Economics* 72, no. 2 (2003): 515–541.
26. Ç. Özden and M. Schiff, eds., *International Migration, Remittances and the Brain Drain* (Washington, DC: World Bank, 2005).
27. S. Carr, I. Kerr, and K. Thorn, "From Global Careers to Talent Flow: Reinterpreting 'Brain Drain,'" *Journal of World Business* 40, no. 4 (2005): 386–398.
28. O. Stark, "Rethinking the Brain Drain," *World Development* 32, no. 1 (2004): 15–22.
29. A. Saxenian, "The Silicon Valley-Hsinchu Connection: Technical Communities and Industrial Upgrading," *Industrial and Corporate Change* 10, no. 4 (2001): 893–920.
30. National Knowledge Commission, *Report to the Nation, 2006–2009* (New Delhi: National Knowledge Commission, Government of India).
31. M. MacGregor, Y. F. Adam, and S. A. Shire, "Diaspora and Development: Lessons from Somaliland," *International Journal of Technology and Globalisation* 4, no. 3 (2008): 238–255.

32. A. Leather et al., "Working Together to Rebuild Health Care in Post-Conflict Somaliland," *Lancet* 368, no. 9541 (2006): 1119–1125.
33. M. Anas and S. Wickremasinghe, "Brain Drain of the Scientific Community of Developing Countries: The Case of Sri Lanka," *Science and Public Policy* 37, no. 5 (2010): 381.
34. N. Fedoroff, "Science Diplomacy in the 21st Century," *Cell* 136, no. 1 (2009): 9–11.
35. E. Chalecki, "Knowledge in Sheep's Clothing: How Science Informs American Diplomacy," *Diplomacy and Statecraft* 19, no. 1 (2008): 1–19.
36. A. H. Zewail, "Science in Diplomacy," *Cell* 141, no. 2 (2010): 204–207.
37. T. Shaw, A. Cooper, and G. Chin, "Emerging Powers and Africa: Implications for/from Global Governance," *Politikon* 36, no. 1 (2009): 27–44.
38. A. de Freitas Barbosa, T. Narciso, and M. Biancalana, "Brazil in Africa: Emerging Power in the Continent?" *Politikon* 36, no. 1 (2009): 59–86.
39. F. El-Baz, "Science Attachés in Embassies," *Science* 329, no. 5987 (July 2, 2010): 13.
40. B. Séguin et al., "Scientific Diasporas as an Option for Brain Drain: Re-circulating Knowledge for Development," *International Journal of Biotechnology* 8, nos. 1–2 (2006): 78–90.
41. R. L. Tung, "Brain Circulation, Diaspora, and International Competitiveness," *European Management Journal* 26, no. 5 (2008): 298–304.
42. House of Commons Science and Technology Committee, *The Use of Science in UK International Development Policy*, Vol. 1 (London: Stationery Office Limited, 2004).
43. T. Yakushiji, "The Potential of Science and Technology Diplomacy," *Asia-Pacific Review* 16, no. 1 (2009): 1–7.
44. National Research Council, *The Fundamental Role of Science and Technology in International Development: An Imperative for the US Agency for International Development* (Washington, DC: National Academies Press, 2006).
45. D. D. Stine, *Science, Technology, and American Diplomacy: Background and Issues for Congress* (Washington, DC: Congressional Research Service, Congress of the United States).
46. K. King, "China's Cooperation in Education and Training with Kenya: A Different Model?" *International Journal of Educational Development* 30, no. 5 (2010): 488–496.

47. R. A. Austen and D. Headrick. "The Role of Technology in the African Past," *African Studies Review* 26, nos. 3–4 (1983): 163–184.
48. S. E. Cozzens, "Distributive Justice in Science and Technology Policy," *Science and Public Policy* 34, no. 2 (2007): 85–94.
49. J. Mugwagwa, "Collaboration in Biotechnology Governance: Why Should African Countries Worry about Those among Them That Are Technologically Weak?" *International Journal of Technology Management and Sustainable Development* 8, no. 3 (2009): 265–279.

Chapter 9

1. P. K. Thornton et al., "Special Variation of Crop Yield Response to Climate Change in East Africa," *Global Environmental Change* 19, no. 1 (2009): 54–65.
2. World Bank, *World Development Report 2010: Development and Climate Change* (Washington, DC: World Bank, 2010), 5.
3. E. Bryan et al., "Adaptation to Climate Change in Ethiopia and South Africa: Options and Constraints," *Environmental Science and Policy* 12, no. 4 (2009): 413–426.
4. "Conditions in Cameroon: A Method to Improve Attainable Crop Yields by Planting Date Adaptation," *Agricultural and Forest Meteorology* 150, no. 9 (2010): 1258–1271.
5. W. N. Adger et al., "Assessment of Adaptation Practices, Options, Constraints and Capacity," in *Climate Change 2007: Impacts, Adaptation and Vulnerability. Contribution of Working Group II to the Fourth Assessment Report of the Intergovernmental Panel on Climate Change*, ed. M. L. Parry (Cambridge: Cambridge University Press, 2007), 869.
6. W. N. Adger et al. "Assessment of Adaptation Practices, Options, Constraints and Capacity," in *Climate Change 2007: Impacts, Adaptation and Vulnerability. Contribution of Working Group II to the Fourth Assessment Report of the Intergovernmental Panel on Climate Change*, ed. M. L. Parry (Cambridge: Cambridge University Press, 2007), 869.
7. M. Burke, D. Lobell, and L. Guarino, "Shifts in African Crop Climates by 2050, and the Implications for Crop Improvement and Genetic Resources Conservation," *Global Environmental Change* 19, no. 3 (2009): 317–325.
8. M. Burke, D. Lobell, and L. Guarino, "Shifts in African Crop Climates by 2050, and the Implications for Crop Improvement

and Genetic Resources Conservation," *Global Environmental Change* 19, no. 3 (2009): 317–325.

9. P. G. Jones and P. K. Thornton, "Croppers to Livestock Keepers: Livelihood Transitions to 2050 in Africa due to Climate Change," *Environmental Science and Policy* 12, no. 4 (2009): 427–437.

10. S. N. Seo and R. Mendelsohn, "An Analysis of Crop Choice: Adapting to Climate Change in South American Farms," *Ecological Economics* 67, no. 1 (2008): 109–116.

11. N. Heller and E. Zavaleta, "Biodiversity Management in the Face of Climate Change: A Review of 22 Years of Recommendations," *Biological Conservation* 142, no. 1 (2009): 14–32.

12. M. Snoussi et al., "Impacts of Sea-level Rise on the Moroccan Coastal Zone: Quantifying Coastal Erosion and Flooding in the Tangier Bay," *Geomorphology* 107, nos. 1–2 (2009): 32–40.

13. M. Koetse and P. Rietveld, "The Impact of Climate Change and Weather on Transport: An Overview of Empirical Findings," *Transportation Research Part D: Transport and Environment* 14, no. 3 (2009): 205–221.

14. National Academy of Engineering, *Grand Challenges for Engineering* (Washington, DC: National Academies Press, 2008).

15. M. Keim, "Building Human Resilience: The Role of Public Health Preparedness and Response as an Adaptation to Climate Change," *American Journal of Preventive Medicine* 35, no. 5 (2008): 508–516.

16. K. Ebi and I. Burton, "Identifying Practical Adaptation Options: An Approach to Address Climate Change-Related Health Risks," *Environmental Science and Policy* 11, no. 4 (2008): 359–369.

17. Y. Pappas and C. Seale, "The Opening Phase of Telemedicine Consultations: An Analysis of Interaction," *Social Science and Medicine* 68, no. 7 (2009): 1229–1237.

18. T. Richmond, "The Current Status and Future Potential of Personalized Diagnostics: Streamlining a Customized Process," *Biotechnology Annual Review* 14 (2008): 411–422.

19. T. Deressa et al., "Determinants of Farmers' Choice of Adaptation Methods to Climate Change in the Nile Basin of Ethiopia," *Global Environmental Change* 19, no. 2 (2009): 248–255.

20. J. Youtie and P. Shapira, "Building an Innovation Hub: A Case Study of the Transformation of University Roles in Regional Technological and Economic Development," *Research Policy* 37, no. 8 (2008): 1188–1204.

21. W. Hong. "Decline of the Center: The Decentralizing Process of Knowledge Transfer of Chinese Universities from 1985 to 2004," *Research Policy* 37, no. 4 (2008): 580–595.

22. S. Coulthard, "Adapting to Environmental Change in Artisanal Fisheries: Insights from a South Indian Lagoon," *Global Environmental Change* 18, no. 3 (2008): 479–489.

23. H. Osbahr et al., "Effective Livelihood Adaptation to Climate Change Disturbance: Scale Dimensions of Practice in Mozambique," *Geoforum* 39, no. 6 (2008): 1951–1964.

24. J. Comenetz and C. Caviedes, "Climate Variability, Political Crises, and Historical Population Displacements in Ethiopia," *Global Environmental Change Part B: Environmental Hazards* 4, no. 4 (2002): 113–127.

25. C. S. Hendrix and S. M. Glaser, "Trends and Triggers: Climate, Climate Change and Civil Conflict in Sub-Saharan Africa," *Political Geography* 26, no. 6 (2007): 695–715.

26. M. Pelling and C. High, "Understanding Adaptation: What Can Social Capital Offer Assessments of Adaptive Capacity?" *Global Environmental Change Part A* 15, no. 4 (2005): 308–319.

27. H. de Coninck et al., "International Technology-Oriented Agreements to Address Climate Change," *Energy Policy* 36, no. 1 (2008): 335–356.

28. S. N. Seo. "Is an Integrated Farm More Resilient Against Climate Change? A Micro-econometric Analysis of Portfolio Diversification in African Agriculture," *Food Policy* 35, no. 1 (2010): 32–40.

29. J. Schwerin and C. Werker, "Learning Innovation Policy Based on Historical Experience," *Structural Change and Economic Dynamics* 14, no. 4 (2003): 385–404.

30. B. Oyelaran-Oyeyinka and L. A. Barclay, "Human Capital and Systems of Innovation in African Development," *African Development Review* 16, no. 1 (2004): 115–138.

INDEX

growth in, 22, 48; electronic wallet (e-wallet) system in, 3; entrepreneurship in, 4–5, 197–201, 203, 206–7, 214–16; fertilizer and, 3, 16; flash drying in, 89; hydropower in, 125; independent power projects (IPPs) in, 126; leadership in, 2, 6–7, 10, 198–99, 265; *Maruca vitrata* (insect) in, 71; national academy of science and technology in, 230; peanut pilot projects in, 161; pea production in, 71; Power Africa initiative in, 127; rice production in, 4; science scholarships for low income students in, 237; sovereign wealth fund in, 6; Staple Crop Processing Zones (SCPZ) in, 6–7; transgenic crops in, 65–66; unemployment in, 216; university industry linkages (UILs) in, 87–89

Nile River, 130

Nile University, 238

North Africa. *See* Middle East and North Africa

Norway, 9

nutrition: agribusiness and, 211; breadfruit and, 211–12; economic-agricultural linkages and, 14; food security and, 23; health care and, 23; innovation and, 10, 13, 26, 151; school gardens and, 158; transgenic crops and, 64, 67–69, 72–73, 78

Obama, Barack, 127

Obasanjo, Olusegun, 2, 198

OECD (Organization for Economic Cooperation and Development), 18

Office du Niger, 129

offices on science, innovation, technology, and engineering, 229–30

oil palm, 4–5, 236

oilseed, 45

One Acre Fund, 205–6

ONE Campaign, 29

OpenCourseware Consortium, 179

Optolab Card project, 56

Organization for Economic Cooperation and Development (OECD), 18

PADRO (Projet d'Appui au Développement Rural de l'Ouémé), 100–101

partnerships: clusters and, 106, 110–11, 113, 115–16; education and, 166, 173, 177, 179; entrepreneurship and, 193–95; infrastructure and, 132, 138, 143–44; innovation and, 110–11, 222–24, 232, 234, 242, 244; international, 33, 42, 106, 222; public-private, 7, 19, 27–30, 32, 110, 113, 115–16, 132, 166, 223

peanuts, 160–61, 164

Peru, 32, 180–82

pest control. *See* insecticides

pharmaceuticals, 250

Philadelphia (Pennsylvania), 243

Philippines, 67, 70–72, 75, 118

PhilRice, 75

PICPE (Presidential Initiative on Cassava Production and Export), 89–90

pilot farmers, 160–61

planting pits, 36, 54

Planting with Peace Program (Nigeria), 7

68; infrastructure and, 118,
120–21, 125; innovation and,
41, 218, 221–22, 228, 245–50;
missions and objectives of
RECs and, xxiii, 34–35
transgenic crops. *See under*
biotechnology
transportation: barriers to, 19;
clusters and, 98; economic-
agricultural linkages and,
16, 19; future and, 257;
infrastructure and, xxi, 117–18,
120–22, 128, 140–42, 144–45,
261; innovation and, 118, 120,
220–22; technology and, 50
Tripartite Free Trade Area (FTA),
245–48
Tunde Group, 100–101
Tunisia, 126
TVEs (township and village
enterprises), 187–89
Twain, Mark, 1

UDS (University for Development
Studies), 155–57
Uganda: aquaculture in, 24;
banana production in, 70–72;
cassava crops in, 79; climate
change and, 36; coffee prices
and, 27; coffee production
in, 174; drought-resistant
crops in, 66, 74; economic-
agricultural linkages in, 27;
education in, 91, 151, 203,
230, 237; entrepreneurship
in, 203; gender inequality in,
119; government-sponsored
agricultural research in, 119;
HIV/AIDS in, 119; horticulture
in, 90–91; infrastructure in,
119, 121, 145; maize production
in, 75; mobile phones in, 50;
national academy of science

and technology in, 230;
rural poverty in, 119; science
scholarships for low income
students in, 237; seed exports
from, 191, 194; service sector
in, 22; technical education and,
25; transgenic crops in, 66, 74;
Water Efficient Maize for Africa
(WEMA) partnership and, 75;
water filtration in, 55; weather
stations in, 136
Ugandan Flower Exporters
Association, 91
Uganda Rural Development and
Training Program (URDT), 151,
153–54, 165
UILs (university-industry
linkages), 87–88, 91, 102–4
United Kingdom: Department for
International Development of,
9, 195, 244; school gardens in,
157; science, technology, and
engineering diplomacy by, 243;
South African wine imports
and, 102; tractors per capita
in, 20; trade and, 102, 196–97;
transgenic crops in, 73
United Nations, 20, 56, 78
United States: drought (2012)
in, 74; EARTH University
and, 168, 171; education
and, 165; innovation and,
242–44; La Molina and, 181;
land-grant universities in,
165; maize production in,
74; nanotechnology and, 55;
National Science Foundation
in, 235; school gardens in,
157; science and innovation
diplomacy by, 242–44;
transgenic crop imports to,
66–67; transgenic crops in,
63–64, 69, 78, 81

United States Agency for
International Development
(USAID): EARTH University
and, 168; Malawi and, 9,
189; science and innovation
diplomacy emphasis at, 244;
transgenic crops and, 71–72;
Victoria Seeds and, 191
Universidad Privada Antenor
Orrego (Peru), 181
universities. *See under* education;
specific universities
universities and research
institutes (URIs), xxi, 83, 110–13
University for Development
Studies (UDS; Ghana), 155–57
university-industry linkages
(UILs), 87–88, 91, 102–4
University of Agriculture
Abeokuta (UNAAB; Nigeria),
88–89
University of British Columbia
Okanagan, 213–14
University of California at
Davis, 75
University of Ibadan (Nigeria), 89
University of Mauritius, 57
University of Pennsylvania, 243
University of Port Harcourt
(Nigeria), 89
University of Stellenbosch (South
Africa), 103
unmanned aerial vehicles (UAVs),
52–54
URDT (Uganda Rural
Development and Training
Program), 151, 153–54, 165

value chains: clusters and, 174;
COMESA and, 249; economic-
agricultural linkages and, 15,
21, 223; entrepreneurship and,
201; infrastructure and, 26;
innovation and, 5; international
trade and, 249; "leapfrogging"

and, 64; Nigeria's Agricultural
Transformation Agenda and,
2; promotion of, xviii; public-
private partnerships and, 223;
regional integration and,
28, 249
Vayalagams, 108–9
venture capital, 94, 172, 176, 185
Victoria Seeds, 191, 194
video technology, 196, 202–3, 258
Vietnam, 233
Vision 2020 (Rwandan
government strategy), 53
Voluntary Guidelines on the
Responsible Governance of
Tenure of Land, Fisheries,
and Forests in the Context of
National Food Security (FAO), 30
Vredeseilanden (VECO), 100–101

Waku Kungo region (Angola),
208–9
wa Mutharika, Bingu, xvi, 2, 8–10
Wang Leyi, 98
WAPP (West African Power Pool),
124–25, 143
water. *See also* irrigation: access
to, 8, 16; Africa's endowment
of, xvi; climate change and,
36; future and, 36; geographic
information systems (GIS)
and, 52; infrastructure and, 6,
42, 128–33, 141–44; innovation
and, 79, 142–43, 224, 228, 251,
257; protocols regarding access
to, 32; purification systems
for, 55–56; regulatory systems
and, 142–43; shortages of, 16,
57, 68, 74, 109, 142, 184, 254, 257;
technology and, 53; transgenic
crops and, 70; Vayalagam
community leaders in India
and, 108–9
Water Efficient Maize for Africa
(WEMA) partnership, 74–75